This inspiring text provides a comprehensive look at the drum kit for what it is – a complex musical instrument. The book covers subjects such as basic acoustical properties to how drum sounds have been modified and presented differently across various genres and eras. The specific instruction on drum tuning is presented in a way that is easy to understand and will benefit beginners, experts, and educators. The online resources available via the companion website are highly valuable as well – I plan to take full advantage of the videos and slides as I discuss drum tuning with my students!
— **Richard King**, Associate Professor, Schulich School of Music of McGill University

Drum Sound and Drum Tuning

Drum Sound and Drum Tuning assists drummers, sound engineers, and music students in learning critical skills related to drum sound and achieving an optimised and personalised drum kit set-up. The book covers the essential theories of percussion acoustics and develops this knowledge in order to facilitate creative approaches to drum tuning and professional-level recording and mixing of drums.

All aspects of drumhead vibration, drumhead equalisation, and resonant drumhead coupling are de-mystified, alongside discussions relating to drumhead types, drum shell vibration, and tuning to musical intervals for different performance genres. The book develops drum sound theory and creative analysis into a detailed dissection of recording and production techniques specifically for drums, including discussions on studio technologies, room acoustics, microphone techniques, phase coherence, and mixing drums with advanced digital audio workstation (DAW) techniques and creative processing tools.

Drum Sound and Drum Tuning includes many practical hands-on exercises that incorporate example tutorials with Logic Pro and iDrumTune Pro software, encouraging the reader to put theory into immediate creative practice and to develop their own listening skills in an informed and reflective manner. The book also documents primary interviews and opinion from some of the world's most celebrated drummers, music producers, and sound engineers, enabling the reader to connect the relevant theories with real-world context, whilst refining their own personalised approach to mastering drum sound.

Rob Toulson is an innovative musician, music producer, sound designer, and studio engineer working across most music and film genres. He has collaborated with many successful music artists and award-winning music producers, including Talvin Singh, Mediaeval Baebes, and Ethan Ash. He is an expert in musical acoustics and digital audio, with a doctorate in vibration and acoustics analysis and a full professor title in commercial music from University of Westminster, London, UK. He is also founder and director of RT60 Ltd, who develop cutting-edge technologies for the audio and music industries.

Audio Engineering Society Presents…

www.aes.org

Editorial Board
Chair: Francis Rumsey, *Logophon Ltd.*
HyunKook Lee, *University of Huddersfield*
Natanya Ford, *University of West England*
Kyle Snyder, *University of Michigan*

Intelligent Music Production
Brecht De Man, Ryan Stables, and Joshua D. Reiss

Women in Audio
Leslie Gaston-Bird

Audio Metering
Measurements, Standards and Practice
Eddy B. Brixen

Classical Recording
A Practical Guide in the Decca Tradition
Caroline Haigh, John Dunkerley and Mark Rogers

The MIDI Manual 4e
A Practical Guide to MIDI within Modern Music Production
David Miles Huber

Digital Audio Forensics Fundamentals
From Capture to Courtroom
James Zjalic

Drum Sound and Drum Tuning
Bridging Science and Creativity
Rob Toulson

For more information about this series, please visit: www.routledge.com/
Audio-Engineering-Society-Presents/book-series/AES

Drum Sound and Drum Tuning
Bridging Science and Creativity

Rob Toulson

LONDON AND NEW YORK

First published 2021
by Routledge
2 Park Square, Milton Park, Abingdon, Oxon OX14 4RN

and by Routledge
52 Vanderbilt Avenue, New York, NY 10017

Routledge is an imprint of the Taylor & Francis Group, an informa business

British Library Cataloguing-in-Publication Data
A catalogue record for this book is available from the British Library

Library of Congress Cataloging-in-Publication Data
Names: Toulson, Rob, author.
Title: Drum sound and drum tuning: bridging science
and creativity / Rob Toulson.
Description: New York: Routledge, 2021. |
Includes bibliographical references and index.
Identifiers: LCCN 2020051474 (print) | LCCN 2020051475 (ebook) |
ISBN 9780367611194 (hardback) | ISBN 9780367611187 (paperback) |
ISBN 9781003104209 (ebook)
Subjects: LCSH: Drum set. | Drum—Tuning. | Drum—Acoustics. |
Popular music—Production and direction.
Classification: LCC MT662 .T68 2021 (print) |
LCC MT662 (ebook) | DDC 786.9/1928—dc23
LC record available at https://lccn.loc.gov/2020051474
LC ebook record available at https://lccn.loc.gov/2020051475

ISBN: 978-0-367-61119-4 (hbk)
ISBN: 978-0-367-61118-7 (pbk)
ISBN: 978-1-003-10420-9 (ebk)

Typeset in Times New Roman
by codeMantra

Contents

Figures

Tables

1 Introduction

A drum kit is a beautiful thing. Perfect geometric-shaped cylinders, grouped together to make something so much greater than the sum of its parts. Simplicity meets design innovation; crafted, wooden shells fixed with precision-engineered metal fittings; rugged hardware stands with glorious polished brass cymbals, pedals and clutches; and an unmissable sparkle finish. Of course, the drums don't need to be made of wood, don't need to have metal hardware, and don't have to have a sparkle finish – but that makes them even more glorious, with infinite possibilities for a drummer to personalise and represent themselves through. Drums are not just beautiful, they are also big, bold, and powerful. We love drums because they are primitive, wild, untamed, loud, and aggressive, yet with the ability to be subtle, simple, and precise. To a non-drummer, or a drummer before they started learning, drums are a mystery. How is it even possible to do four different things with four different limbs? Of course, we drummers know that our limbs do not perform independently, but are all connected together by a musical system in our brains, which enables each component of our body to contribute collectively to a single instrument sound – the sound of the drum kit.

Alex Reeves, drummer for Elbow and a number of other esteemed artists (including Bat For Lashes, Anna Calvi, and Dizzee Rascal), gives a valuable insight into the beautiful and complex world of drum sound for performance and studio recording:

> Sometimes a drum just has a magic or "thing" about it that sounds great in a certain context or room, sometimes it sounds a bit like a biscuit tin to your ears but the mics just love it! However, there are some universally recognised great sounds for drums: rich and full low-end for bass drums, a bit of bite in the high-mids for snares, overall not too much of the low-mid presence that can get in the way of guitars and vocals. The way you hit em and tune em can make all the difference – when recording I'll sit behind the kit with the song playing in my headphones changing snare drums and mics until we've got the right overall character, then micro-tune, dampen if necessary, changing the tension

of the snare wires, changing where and how hard I hit the drum. Just amplitude and mic gain can make such a massive difference to the overall sound of the kit. But some drums are just "right"! Often it is the more beautifully-made or classic drums that get onto a record, but occasionally it's the grotty, nasty sounds that give the character – not everything has to be pristine.[1]

Through the course of this book, we'll unpick all of these creative and technical concepts with respect to popular drums and give a complete hands-on approach for becoming competent and knowledgeable with drum sound and drum tuning, in context for both drummers and sound engineers.

1.1 Values of great drum sound

Drums occupy a very important and loved aspect of the sound spectrum that we hear, which we can refer to as *low-frequency* (or *bass*) sound. Most people will understand this term and identify the difference between *treble* and bass and the similar settings used on a hi-fi or amplifier system. And most will have felt the boom of a kick drum resonating in their chest at a nightclub or concert venue, or the powerful low-frequency drumrolls around the tom drums on classic rock and metal tracks. Low-frequency is obviously a relative term, and while we will go into concepts of frequency and the sound spectrum in much more detail in the book, it's very easy to identify the difference between a low-frequency sound such as a kick drum, tom drums, or a bass guitar, and a higher frequency sound that we would hear from the top keys on a piano, from a flute, a mandolin, or someone singing the high notes in a scale. Drums, of course, have high frequencies too, from the overtones of a ringing tom, the wires interacting on the underside of a snare drum, the crack of a hard rimshot, and the shimmer of the cymbals. But nothing else really delivers or owns the low-frequency range in a popular music track as much as the drums. In much popular music, bass guitar, double bass, or a bass synthesizer is also a charismatic musical element in the low-frequency range. But, in many cases, these instruments are still performed in companion with the drums and often have a role that also gives a percussive power and presence to the music. So the drums are extremely important in performed and recorded popular music, and this is why professional musicians and music producers put considerable effort into getting the best possible sonic characteristic from a drum kit.

There are a huge number of great sounding drummers and drum recordings, for example, John Bonham of Led Zeppelin is one of the most discussed specifically in terms of drum sound. Bonham set up his kit to give distinct resonant pitches for each drum and worked with the band's studio engineers and music producers to create a big reverberant sound to the drum kit. Conversely, Ringo Starr and George Martin were particularly well known for keeping the drum sounds on *The Beatles'* recordings very dry and with minimal reverberation added.[2] So it's clear there's no single approach to drum sound that is most correct. It's great that every drum kit looks different, but, more importantly, every drum kit also sounds different. If there are a million different styles, constructions, and visual designs of drum kits, then there are a billion different sounds of drum kits. Alongside learning to play drums, every drummer embarks on a journey to develop the sound and set-up of their kit, which can take many years and some will claim never ends (and never should!). This can be quite a personal journey of exploration, since what is right for one drummer may be totally wrong for another, as emphasised by drummer Cherisse Osei (drummer for Simple Minds, Paloma Faith, and Bryan Ferry):

> What sounds good to me might not sound good for somebody else, as it's a very personal thing. If it doesn't resonate with me or I don't get excited by the sound, then I know it's not for me.[3]

So the modern drummer can choose from an almost infinite number of permutations in their set-up, which means no two drum set-ups could ever look or sound the same. Do you want a 3-, 4-, 5-, or 15-piece drum kit? What shell material gives the sound you are looking for? What drumheads do you like the sound of on your kit? How many cymbals do you need? Do you want a double kick drum pedal? Sticks, brushes, or mallets? Do you like them tuned low and powerful or high and resonant?

That last question brings us to an important point: tuning. All the prior questions relate to the design and set-up of the kit – deciding how many drums, what kind and by which manufacture, and which accessories and replaceable parts you prefer to use with the kit. But the question about tuning implies that, even once you have your perfect drum kit set-up and positioned in the studio, performance stage, or your spare room, there are still multiple options available with respect to how the kit is tuned and how the kit can sound. When added together, this gives both a wonderful sweetshop of possibilities and at the same time a treacherous minefield of unknowns, hazards, and challenges.

1.2 Why bother with drum tuning?

It's a good question and one most experienced drummers will be very aware of the numerous answers to. Drums and drummers often receive a hard time from other musicians, because they are so "straightforward" in comparison with many instruments – you just hit them and they make a noise! But, in reality, drums are much more difficult to tune than a guitar or violin, for example. Guitars and violins have one single tuning head for each string, which is turned until you reach the exact musical note you are looking for. You might not have perfect pitch hearing, but it's possible to listen to a piano until the string and the piano note you are looking for sound the same, or use an electronic tuning device that indicates exactly when the correct note is reached. Now that's simple!

But drums have a whole series of different things to consider. Not least, there are potentially six, eight, or ten tuning rods on each drumhead, which can all individually affect the sound. But this arrangement is doubled, because there's a second drumhead on the bottom, which also has six, eight, or ten tuning rods. In addition, there are a number of drums in the kit, each of different sizes and depths and a multitude of drumhead styles to choose from, and, furthermore, each song and music genre potentially requires a slightly different drum sound. Before you know it, there are just too many permutations to make sense of! And so tuning drums can be a difficult task, though many experienced drummers get really good at it over the years. But we haven't really answered *why* bother to tune drums yet; there are lots of reasons:

First and foremost, tuning can help to get the absolute best sound out of your drums and drum kit. If your drumheads are too slack or too tight, then the drum won't resonate and project sound as well as it can. If a drumhead is tuned unevenly (e.g. tight on one side and loose on the other), the drumhead won't vibrate smoothly and will choke the sound or introduce warbled artefacts to the sound. It's amazing what a well-tuned drumkit can do for a drummer – it can often be more enjoyable to hear an average drummer play simple on a wonderfully tuned kit, than to hear an amazing drummer hitting sublime grooves on an awful sounding set of drums!

Secondly, it's valuable to find the perfect sound for your drums that is best for the songs you are playing, the music genre, and most importantly your own personal style and character. It's not uncommon for jazz drummers to tune their drums really tight and high so that musical notes ring out from their toms and they can play musical phrases around the kit. Conversely, rock drummers tend to tune quite low and powerful to bring a deep tone and style to the performance. With this in mind, it's quite possible to suggest ranges of tuning frequencies for different styles and kit set-ups, but there are no fixed rules, and you can develop your own style too. If you can play high-tuned toms in a melodic rock band, then go for it, and the chances are you'll be noticed for doing something different too!

Additionally, like all instruments, it's important to be consistent – to sound the same tomorrow as you do today, and to be in harmony with the other instruments and the song you are playing. We use tuning of instruments to take a record of our settings and then remember for the next time. It's easy with guitar and other instruments, because if a string is tuned to a specific note (E, A, G, etc.), it's quite easy to remember. But drums don't have to be tuned to exact musical notes, so there are a multitude of possibilities for how you might tune your drums, and that's why it's much harder to identify exactly what your previous sound was, especially if you've just changed drumheads or are on tour moving kit around day and night.

Having said that, it is actually possible to tune drums to musical pitches, and this can sometimes be valuable for songs that have big tom fills that need to be harmonious with the bass guitar or synth line. Some drummers even use a piano to help them tune their drums to specific notes. Equally, tuning to musical pitches can help with deciding on the different tunings for each drum in your kit; so if each is tuned to a different musical frequency, then you can be sure a fill around the kit or a repeating pattern through the toms will sound pretty cool!

While we're talking about recording settings, drum tuning is perhaps most important when in the recording studio, because it is there that a sound will be committed to a record that will hopefully be heard by audiences all over the world and for many years to come. Getting a great drum sound is essential in many recording sessions, and maintaining consistency through different takes and days, and when changing drumheads, can be really important. Great drums can really define a record, and legendary music producer John Leckie (Muse, Radiohead, and The Stone Roses) is quoted as saying:

The two things that identify a record are the vocal and the snare drum[4]

So producers tend to spend a long time achieving the right drum sound before starting a recording session, where possible taking the first day or two just focusing on the drums, tuning, and microphone placement, which can all potentially account for 15–25% of the recording session.[5] These days, record companies are no longer willing to spend large funds on recording projects, so a "right first time" approach makes good economic sense. If the drum set-up is correct, the recording session can move swiftly and efficiently, which can also help the mixdown process too.

Hopefully you agree, playing drums well is really important for drummers, but it's a huge asset if you can tune drums and get a great sound out of them too. This is equally applicable for studio engineers who regularly record drums too. It's clear that beginners really need help learning about the concepts of drum tuning, putting their knowledge into practice and developing the skill quickly – it takes a long time to learn and to develop your hearing to the point where you can tune without any help or feedback of some kind. But it's also the case that professional drummers and drum technicians often need to tune quickly, repeatably, to an extreme level of consistency between sessions (e.g. on a big tour where the sound must be the same every night). Furthermore, our ears are only so good. We humans can sometimes only hear to an accuracy of a few hertz (or Hz), and yet drums tuned to higher accuracy than this can have a positive influence on the drum sound. In particular, very precise tuning is beneficial when tuning drums to specific musical frequencies or when trying to equalise a drumhead to perfect consistency.

1.3 About this book

This book is intended as a complete and holistic resource for anyone wishing to learn more about drum sound, in order to get great results in setting up a drum kit for performance or studio recording. It is therefore aimed at drummers and sound engineers who are embarking on a lifelong journey with the drums. It is suitable for drummers and engineers of all abilities, with the aim of connecting the intrinsic knowledge of experienced musicians to the core theory and science of drums, which will unlock their creativity and success with drum performance and recording. This book will also help new drummers and sound engineers, and students on music, acoustics, and music production courses, by discussing the fundamental knowledge areas related to drum sound, which will set them on an empowered and informed career in the music industries.

This book covers three key topic areas in a linear order, to enable each chapter to reflect and build on the concepts covered in previous chapters. The first chapters discuss the acoustic principles of drums, from drumhead vibration concepts, the relationships between the two drumheads, and the fundamental theories which give us control over the sound of drums. Armed with this knowledge, readers are then equipped to embark on the practical aspects of drum sound and drum tuning, understanding how to achieve a great sound of any drum with a simple holistic approach, and further discussing the role of drumhead design and shell vibration in the sound of drums. Finally, the book discusses theory and techniques related to recording, mixing, and producing drums to a high professional standard, relating back to the theories of drum sound in order to create recordings of drums that complement the music genre, have character and impact, and, where relevant, are also larger than life!

1.3.1 Educational approach

The book takes a structured educational approach that allows you to learn and implement the discussed topics, in order to lock in new skills in a practical context. It's true that you can only learn so much from a book, article, or online video; deep learning only really happens through practical implementation, so you are encouraged in each chapter to implement the techniques and theories with your own drum kit or studio resources. Reflective practice is essential too, in order to master a difficult skill, so you'll need to listen, think, experiment, and evaluate the outcomes in order to excel and fully understand the written content. To encourage this, you'll see numerous "Try for yourself" exercises through the book, which will make all the difference in bringing the discussed theories to life. If you are a course tutor, then these practical exercises can easily be adapted to support a hands-on course in drum sound, drum acoustics, and drum recording and mixing. At all stages, the book is focused on connecting the science and art of drum sound. This enables drummers and sound engineers to draw together two complex subject areas, which are much more powerful when combined and considered in tandem. Creativity always pushes science to its limits, and new scientific knowledge empowers creativity through the course of technical evolution!

The discussed theories are also verified with respect to the most cutting-edge research knowledge on the subject of drum acoustics, as well as with reference to some of the most esteemed drummers, studio engineers, and authors in the field. You'll see that each chapter has a list of endnotes, which verifies the validity of the content in the book and allows you to check some of the discussed topics in more detail with further reading. The subject of drums is a huge one, and some discussions extend outside the scope of this book, but you'll always be pointed in the right direction to delve deeper and research into parallel discussion topics if you wish.

1.3.2 Online resources

An online companion website supports the content in the book, including open-access blog posts, tutorial videos, and links to high-resolution colour images of all the figures in this book. The online webpage also includes access to prepared presentation slides for each chapter, enabling a course tutor to quickly prepare course material related to drum acoustics, drum tuning, and drum production. The book's companion website can be found here: www.drumacoustics.com

1.3.3 Interviews with esteemed professionals

In writing this book, a number of esteemed musicians and music producers were asked to give their thoughts and opinions on drum sound, for inclusion in the book and to help connect the discussed theories with practical and expert practice in the music industries. It is with great honour that these quotes can be included as primary reference points, and you may wish to research the interview contributors further. Primary quotes from the following professional practitioners are included:

MIKE EXETER: Grammy winning music producer and studio engineer for Black Sabbath and Judas Priest.
SIMON GOGERLY: Grammy winning producer and mix engineer for U2, No Doubt, and Underworld.
SYLVIA MASSY: Grammy winning music producer and studio engineer for Tool, Taylor Hawkins, and System of a Down.
CHERISSE OSEI: Drummer for Simple Minds, Paloma Faith, and Bryan Ferry.
EMRE RAMAZANOGLU: Drummer and award winning music producer for Noel Gallagher, Kylie Minogue, and Richard Ashcroft.
ALEX REEVES: Drummer for Elbow and many other artists including Bat for Lashes, Anna Calvi, and Dizzee Rascal.

Full transcripts and more detailed interview responses from the expert contributors can also be accessed from the book's companion website.

1.3.4 Companion software and examples

Many examples in this book refer to two particular software applications: iDrumTune Pro and Logic Pro. While it is not essential to own either of these applications, there is certainly a benefit to be gained by implementing some of the examples shown in this book and exploring the discussed concepts in a practical context, which these software applications support.

The mobile phone app iDrumTune Pro is developed from the novel research of the book author Professor Rob Toulson and is available on iPhone and Android platforms from the respective Apple and Google Play app stores. The iDrumTune Pro application allows a quantitative analysis of drum acoustics and drumhead vibration, meaning that the acoustics theories discussed in the book can immediately be verified and experimented with by the reader. The iDrumTune Pro app was first developed as a novel research tool to assist with gathering new acoustics knowledge on drums and drumheads,[6] but in recent years, it has become a valuable and well-respected app for assisting drummers and sound engineers in learning the concepts of drum tuning and achieving accurate and repeatable drum set-ups.

Logic Pro is a well-known *digital audio workstation (DAW)* package that is used by many musicians and sound engineers, from beginner to professional standard. This book uses Logic Pro to exemplify some of the acoustics concepts relating to vibration and sound theory, and also provides a platform to discuss set-ups and layouts with respect to drum recording and mixing drum sounds. Whilst examples and screenshots are shown specifically from Logic Pro, it is very possible to set up equivalent examples and experiments with any modern DAW software package. In addition to Logic Pro, a number of third-party effects *plug-ins* used for mixing drums will be mentioned too towards the end of this book.

1.4 Don't forget to listen!

Whilst this book is concerned with discussing the theories and techniques associated with drum sound, drum tuning, and recording studio production of drums, it is a fundamental aim of the book to help you develop your listening skills with respect to drum sound and drum tuning. Software tools and acoustics theories give guidance to follow and allow critical feedback on your abilities, which are fundamental to learning. But, if you follow this book and implement the practical exercises, an implicit by-product will be the development of your hearing, to the point where you may be able to set up drums and tune them to high precision unassisted. In fact we'll see in this book that the human ear is not so readily designed for drum tuning, and learning to develop your hearing with respect to drum sound is not so easy without an understanding of the sonic characteristics to listen for and the musical acoustics theories that they relate to. With this in mind, by developing your knowledge, practical skills, and hearing abilities, this book will help you become a master of drum sound, which will support a whole lifetime of enjoyment with drums and percussion instruments.

Notes

1 Interview with drummer Alex Reeves conducted on 07/10/2020.
2 Interview with drummer Cherisse Osei conducted on 27/10/2020.
3 Matt Brennan gives an excellent and detailed account of the history of drums, performance techniques, and the sounds which different drummers have aimed for, including discussion of John Bonham and Ringo Starr, in *Kick It: A Social History of the Drum Kit* by Matt Brennan, Oxford University Press, 2020.
4 John Leckie quoted in *Behind the Glass: Top Record Producers Tell How They Craft the Hits* by Howard Massey, Backbeat Books, 2003, p. 104.
5 As discussed by Toulson et al. in *The Perception and Importance of Drum Tuning in Live Performance and Music Production*, Proceedings of The Art of Record Production Conference, Lowell, Massachusetts, November 2008.
6 *Tuning or Training Device*, UK Patent Application by Rob Toulson and Anglia Ruskin University, Number GB0713649A, 2007.

2 Drumhead vibration and the science of sound

Drums generate sound through vibration, but what actually is vibration, how does that relate to the sound we hear when a drum is hit, and what does that have to do with drum tuning?

We call the analysis of sound vibration *acoustics*, which is a valuable topic area for any musician or studio engineer to have a basic grasp of. This is because, as musicians, we are responsible for our own instruments and the quality of the sound they create. If we don't understand that old guitar strings or broken drumheads fail to vibrate properly, then we wouldn't know that their sound is compromised and some maintenance is required. This is particularly relevant for drummers, since there are so many ways to customise and personalise a drum kit set-up. Without a little knowledge of sound and how it is created and controlled, it would be impossible to decide what to modify or change in a drum set-up, or which drumheads to use, or whether the expensive solid oak snare you like the look of will actually give the sound you are aiming for. Moreover, with a little understanding of musical acoustics, you'll be able to unlock many creative opportunities in how you set up and play your drumkit. Acoustics theories explain why a rimshot sounds different to a ghost note, and why some rooms make your kit sound great and others seem to generate a sonic mushy mess of sounds competing and washing over each other. If you have a fundamental grasp of instrument acoustics, then you will be well equipped to get the best out of your equipment and to quickly resolve any issues that arise during your travels and exploits as a musician or sound engineer.

2.1 Sound sources, acoustic transmission, and sound reception

Vibration is quantified by the number of times something moves backwards and forwards in a single second, i.e. its *frequency*. The measurement unit for vibration frequency is known as *hertz* or *Hz*. So, if a drum is tuned to 100 Hz, its drumhead will vibrate up and down 100 times in a single second. This is obviously too fast for us to see, but the vibration really does happen with a very small movement. So, when it is hit, the drumhead vibrates, and that vibration causes a very small change in pressure of the air molecules

Sound Source   **Transmission**   **Receiver**

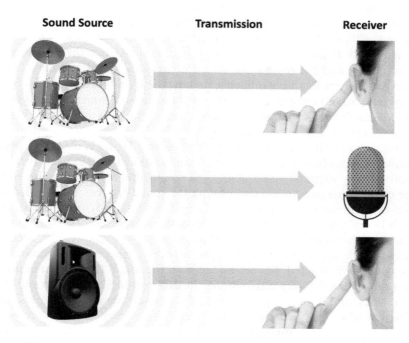

Figure 2.1 Sound sources, transmission through air and reception.

around it. The pressure disturbances also vibrate and transfer through the air to our ears or a microphone, so we can hear and record the information from a drum or any other sound source. Figure 2.1 shows some fundamental examples of sound creation, transmission, and reception.

The sounds we hear in life are all made up of complex acoustic interactions, influenced by the source of the sound, the transmission of the sound through air (usually, but sound can travel through solids and liquids too), and the reception of sound at our ears or a sound recording device. The actual sound that is experienced or captured depends on all three of these things, and there can be huge variations in reality, which is why no two things ever really sound 100% identical. For example, sound sources can be absolutely anything, but in musical terms a sound source can be hugely varied, as you know; two 12″ rack toms can look quite similar but give off very different sounds depending on how they are constructed, tuned, and performed. Considering acoustic transmission, the same drum will sound very different depending on the environment it is in; you'll know that if you take one of your drums into the kitchen, it will sound very different than if you take it outside into the yard. The third component is the sound receiver, usually our ears or a microphone. There is huge variation with this too. Not only do everyone's ears convert sound to brain information subtly differently (we all have different sized ears after all!), but our brains all interpret sound slightly differently too. Our hearing changes with age and based on how tired or excited we are, and we can train our hearing too. We'll see later on in the book that it's actually possible to trick or confuse our ears and hearing too. If the sound receiver is a microphone, then the design of the microphone and the method by which it captures sound and coverts it to electrical energy has a big influence on how the sound is recorded for future playback. So, all components of sound transmission from source to reception define the sonic characteristics of a sound, and it is no surprise that we have sensitive hearing that can identify subtle differences between things.

Referring to the lower two images in Figure 2.1, and considering the example of recording and playing back the sound of a drum kit, there are clearly many aspects which define the sound that we hear at the end of the chain. First, we have a drumkit as the sound source, and we all know that every drum kit sounds quite different, and every drummer achieves a slightly different sound with their performance style too. The sound of the drums is transmitted through air, and the way the sound travels, incorporating all the sound reflections and pressure disturbances in the room where the drums are situated, has an influence on the sound that will be captured by a microphone. Every microphone has different internal design and characteristics, and it's no surprise that different microphones capture sound in different ways, and hence affect the sound themselves. When the recorded sound is played back through a loudspeaker, its characteristics are affected again, given that different speakers have different designs and very different sonic qualities. The sound played back is also propagated through a room or some kind of space, so the sound is again affected by the transmission medium. Finally, the sound reaches our ears, where the final conversion into brain waves or sound information that we interpret takes place, which is very different for all humans, since two people can hear the same sound and respond very differently to it. So, the sound received by the ear, in this example, is potentially very different from the sound that is actually made by the drum kit at the start of the chain. It still sounds like a drum kit, but it is modified by the transmission medium and the technology used to record it and play back, and it is interpreted and perceived uniquely by our brains too. As musicians and sound engineers, it's valuable to have an understanding of all these aspects of sound creation, transmission, and reception, in order to ensure that your listeners hear and interpret your performance or recorded sounds in the way that you would like them to!

2.2 Evaluating frequencies

In the world around us, large objects tend to vibrate more slowly than small objects. Tall buildings sway from side to side very gently every few seconds and large lorries have heavy suspension systems that bounce up and down in response to bumps in the road. In comparison, a dinner knife from the cutlery draw vibrates thousands of times per second if you hold it at the end and hit it on a hard surface, emitting an audible "ting" sound as it moves. Some things don't vibrate at all or stop vibrating very quickly, such as a cushion or a rubber toy – if you hit these types of things, they make a dull thud sound, but don't really vibrate after the initial impact has completed.

We can't usually hear the types of vibration associated with buildings shaking and lorry suspension springs moving up and down. As humans we can only hear vibrations that occur really quite fast, like the knife from the cutlery drawer when it is hit on the tabletop. The range of human hearing is between about 20 and 20,000 Hz (vibrating 20 times per second to 20,000 times per second), which is quite a big difference. The topic of acoustics extends into understanding how these frequencies relate to each other and how audible frequencies can be manipulated and organised to sound pleasant and musical.

Since larger things vibrate slower and at lower frequencies than smaller objects, it should be no surprise that a large diameter drum gives us a low-frequency sound, and a small diameter drum generally has a higher frequency. This applies equally to guitar strings, where the thinner strings give higher vibration frequencies. In musical terms, we refer to the strongest vibration frequency as the *pitch* of the note being played. The tension of a string or drumhead also affects frequency. The tighter the drumhead, the higher the vibration frequency, as applies also to guitar strings. So, by tightening or loosening the *tension rods* positioned at *lugs* mounted around the outside of drum, we can increase and decrease the frequency to achieve our preferred sound or pitch. Correctly, we apply *torque* to tension rods (or *tuning rods*), which means applying force in a circular motion, though with drums we often use the term *applying tension* to refer to tightening the rods; this phrase makes sense because the result of applying torque to the rods alters the tension of the drumhead, in the same way that applying torque to a guitar machine head changes the tension of a guitar string. In very general terms, we use the word *tuning* to refer to the set-up and optimisation of any system (we can *tune* an engine, an algorithm, and a musical instrument!). So, we generally use the word *tuning* quite liberally with respect to drums, referring to all aspects of optimising the sound of a particular drum kit, not just applying torque to the tension rods. We'll be considering all of these aspects of drums and drum kits which can be optimised, tuned, and manipulated in much more detail through this book.

TRY FOR YOURSELF: EXPERIENCE FREQUENCIES!

It's quite a simple thing that we don't really do that often, but as a percussionist it's valuable to have an understanding of how things vibrate and how the characteristics of that vibration can be identified in the resultant sound. Take a drumstick or mallet and go around your house gently tapping things. Look for household items which give off more

(Continued)

musical sounds than others, particularly those that vibrate at interesting low or high frequencies, those which seem to have richer sounds, those which ring out for a long time, and those which stop quickly. Also have a listen to how the sound of things changes depending on where you tap them or how/where you hold them, what you tap them with, and what room you are in. For example, try tapping a mug, the backrest of a chair, a child's toy, a radiator, taps and the bath tub, or anything else that might make an interesting sound.

2.3 The single most valuable musical acoustics theory!

There's one key acoustics theory that is absolutely fundamental to drum tuning. Understanding this simple theory can immediately improve your knowledge and ability with drum tuning, and as a result, you will learn the skill of drum tuning faster, with more knowledge of the instrument, and you'll find it much easier to get the sound you want from your drum kit. You'll be able to tune more precisely and more consistently, and, if you wish, to musical notes and intervals on the drum kit. It's quite simple; basically the drumhead vibrates at many different frequencies all at the same time. This acoustics theory applies to all musical instruments, but is particularly relevant for drums and drum tuning. So when you hit the drumhead, many different frequencies are heard as it attempts to vibrate in different ways and shapes. We call these frequencies and vibration shapes *vibration modes*, and depending on where you hit the drum, some modes and frequencies resonate stronger and louder than others.

Two of these frequencies and vibration modes are the most important for drum tuning, and it's really easy to hear the difference in them, even if we can't see the drumhead vibration shapes themselves. The two key drumhead frequencies are as follows:

- The *fundamental* frequency, which we will refer to as *F0*, is excited most when the drum is hit at the middle and sounds like a "BOOM."
- The *first overtone* frequency of the drum, which we will refer to as *F1*, is excited most when the drum is hit at the edge and sounds like a "PING."

It's really easy to hear the difference between these frequencies: just by hitting between the middle and the edge repeatedly, you'll hear the BOOM at the centre and the PING at the edge. This applies to all musical instruments – if you take a guitar and strum it in the middle of the strings, the sound is warm and full; then if you strum closer to the bridge, you'll hear more overtones and the sound becomes quite thin with stronger treble frequencies. It's exactly the same concept as hitting the drum in the middle or at the edge.

So, what's actually happening here, and why does the drumhead vibrate differently and sound different when it is hit in different places? Well, the reason the drumhead can give different sounds is because there are many different ways a vibrating drumhead can physically deform and vibrate. In simple terms, the drumhead can vibrate on a circular axis (around the drumhead) or in a diagonal axis (across the diameter of the drumhead). Figure 2.2 shows the fundamental F0 vibration mode and its circular axis, as well as the F1 overtone mode and its diagonal axis. Indeed, we refer to the fundamental as F0 because it does not have a diagonal vibration axis, and we refer to the first overtone as F1 because it has exactly one diagonal vibration axis.[1] The F1 mode actually has the circular axis too, since the drumhead can't move at all around the edge of the drum.

The reason the drumhead sounds different when hit in the two different places is because the fundamental and overtone modes have their biggest vibration movement or amplitude at different positions on the drumhead. The F0 vibration shape has its biggest amplitude in the very centre of the drum, so if we hit the drum in the centre, we excite this mode significantly. Conversely, the F1 vibration mode has no amplitude at all at the centre (it moves up and down either side of the diagonal centre line), so it is not excited as much with a drumstick hit in the centre. The opposite applies nearer the edge of the drumhead; the F1 mode has its greatest amplitude towards the edge, and the F0 mode has much less amplitude towards the edge. So, an edge hit causes the F1 frequency to be more distinctive than the F0 frequency. This explanation also applies to a microphone's position when recording a drum: If a microphone is positioned at the centre of the drumhead, where there is no movement of the F1 mode, much less F1 frequency will be captured into the microphone. If the microphone is positioned towards the edge of the drumhead, then there is a greater amplitude of the F1 vibration frequency to be captured, and less of the F0 frequency. This explains why the microphone positioning on any musical instrument (or loudspeaker) can make a huge difference to the sound captured.

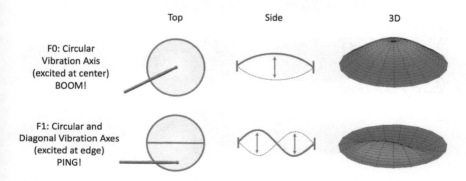

Figure 2.2 Fundamental and first overtone vibration modes of a drumhead.

We use the F0 fundamental (centre frequency) to set the overall pitch of the drum. This can be high for jazz-type drumming or low for rock-style drumming, and anywhere in between too. Every drum has a range between the drumhead being too slack and the drumhead being too tight, so there are lots of frequencies and tunings to experiment with, based on the type of music you are playing and your personal style too. If you identify and remember what F0 frequency you prefer for each of your drums, then you can make sure your drum is always tuned to this frequency every time you play. Furthermore, if you play in different bands, you might use different tunings for different music genres, so understanding drumhead frequencies helps you quickly get the sound you need at any moment in time and can be useful when changing drumheads or to make a record of the tunings used in a recording session.

The F1 overtone frequency, which is most prominent when hitting the drum at the edge, is important to ensure that the drum is tuned evenly and vibrates with a clean and consistent profile. If one point around the drum has a slightly different F1 frequency, this causes a vibration clash on the drumhead which makes the sound of the drum warble or modulate. In acoustics, we call this condition *beating*, which occurs when two or more close but not exact frequencies are present at the same time. Beating in the drumhead means that the sound is not smooth, and you don't get a pure tone of your drums when they are hit.

In fact, there aren't just two vibration modes, but hundreds of modes and frequencies, and this principle applies to all musical instruments and vibrating objects too. The more complex modes are identified by the fact that the drumhead also vibrates with two and more circular axes, and with more than

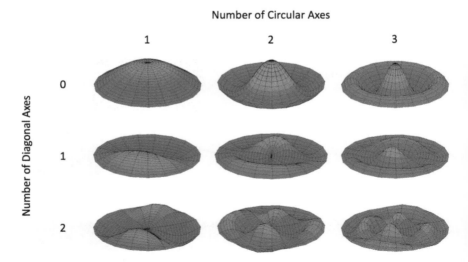

Figure 2.3 Drumhead vibration modes with different numbers of circular and diagonal axes.[3]

one diagonal axis too, and combinations of all the different numbers of circular and diagonal axes also. You don't need to worry much about these extra vibration modes and frequencies, but they are all evident in a vibrating drumhead. The balance and relative power of these frequencies is what makes one drumhead type sound different to another, because different drumhead thicknesses, deigns, and materials cause different amounts of each vibration mode to be excited when the drumhead is hit, giving each drumhead a unique sound and character. Figure 2.3 shows nine different vibration modes for a single drumhead. It's not really possible to see these modes by eye, but with advanced acoustics measurement tools, it is possible to prove that they are all there.[2]

2.4 Measuring and analysing drum modes

In audio and acoustics, we often use what's called a *spectrum analyzer* to see a graph of all the different frequencies that are heard in a sound signal. The iDrumTune Pro app has a spectrum analyzer feature built in, so it's possible to see what vibration frequencies are being immitted by a drum, and it's therefore possible to be quite certain which is the F0 fundamental and which is the F1 overtone.

For example, if a drum is hit in the center, we expect F0 to be excited most, which is seen as a big spike on the spectrum analyzer at the F0 frequency (Figure 2.4a). If the drumhead is hit at its edge, the first overtone is excited most, so we see the spike in a different place at a higher frequency (Figure 2.4b). What's interesting is that if the drum is hit somewhere between the centre and the edge, we excite both at the same time (Figure 2.4c). When listening, our ears might not always be able to pick out or distinguish between the two frequencies, but the spectrum analyzer is more than capable of showing us what is happening in terms of drumhead vibration. It's no surprise to see that there are a few other much smaller spikes too on the frequency spectrum, representing the other vibration modes of the drum that are less powerful. Generally, for tuning, we can ignore the exact values of

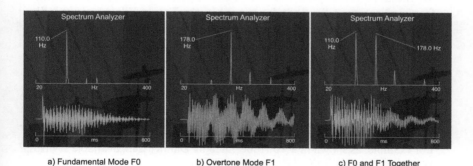

a) Fundamental Mode F0   b) Overtone Mode F1   c) F0 and F1 Together

Figure 2.4 F0 and F1 vibration modes identified by the iDrumTune Spectrum Analyzer.

these other vibration modes, as the F0 and F1 frequencies tell us everything we need to know in order to tune a drum so that it sounds smooth, tuneful, and powerful.

TRY FOR YOURSELF: IDENTIFYING A DRUM'S FUNDAMENTAL AND OVERTONE FREQUENCIES

Focusing on the tom drums in your kit, listen to the difference between the F0 and F1 sounds of each drum. Just tap the drum gently at the centre four times and then at the edge four times. Then try with slightly more force and speed on the hits, and you should clearly be able to identify the BOOM and PING sounds on each of your drums.

Now use the iDrumTune Spectrum Analyzer feature to measure these frequencies on your tom drums. Hold the device microphone just a few centimetres away from the hit position each time, and you should be able to see a prominent low-frequency spike for centre hits and readings, and a higher frequency spike for edge hits and readings.

Now hit one of your tom drums at the center and take a reading with iDrumTune. Gradually move further and further away from the centre with each reading, and you should see the F0 spike start to reduce and the F1 spike start to increase as you move towards the edge. Can you find a place on the drumhead where they are both excited about the same amount? Can you hear the sound of the F0 and F1 frequencies at the same time and distinguish between them?

The iDrumTune app also has a Target Filter feature, which is designed to help out when you want to read F0 and ignore F1, or vice versa. It's worth getting an understanding of how this works and how it can be useful in drum tuning. Take a reading from a drum that excites both the F0 and F1 frequencies, and then press the Target Filter button a few times to see how it is possible to select a specific frequency range for the analysis. If you find that a drum gives off many very strong frequencies, regardless of where it is hit, the Target Filter feature will allow you to focus on a frequency range close to F0 or F1, as desired.

Notes

1 In vibration theory, we also use the terms *nodal circles* and *nodal diameters* to describe the vibration axes of a circular membrane. This theory is discussed in more detail in *Science of Percussion Instruments* by Thomas D. Rossing, 2000, Volume 3, World Scientific Publishing Co., p. 6.

2 For example, as shown in a YouTube video by Dan Russell, available at https://youtu.be/v4ELxKKT5Rw [accessed 01/08/2020].

3 Animations of circular membrane vibration modes can be generated by mathematical implementation of Bessel functions. Some example animations are shown in a YouTube video by Spiros Kabasakalis, available at https://youtu.be/DZ8VGAx4178 [accessed 01/08/2020].

3 Tuning the pitch of a cylindrical drum

Cylindrical drums are those us drummers are most used to playing - cylindrical shells with two drumheads, one on each of the batter (top) and resonant (bottom) sides. No matter how they are tuned, all cylindrical drums have a sonic pitch, which in very simple terms means the frequency that the drum vibrates at and which we hear with the most prominence. If you make the drumhead tighter, the pitch goes up; if looser, the pitch goes down. It's no different to a guitar or violin – if you tighten a string, the sound goes up in pitch; if you loosen a string, its pitch goes down.

Musical pitch is an interesting concept related closely to vibration frequencies, the difference being that the pitch is a very human quality, relating to how we hear or perceive the different frequencies. It's a subtle difference, but our ears can sometimes be tricked into perceiving a different pitch to that which is objectively measured by an accurate acoustic analyzer. Nevertheless, in many cases, musical pitch and the fundamental frequency that is created by an instrument can be considered to be equivalent. It's commonly accepted that humans can perceive the pitch of a sound if it has "periodic acoustic pressure variations", which is virtually anything that vibrates backwards and forwards for more than a few cycles[1]; this includes guitar strings, xylophone bars, and of course drumheads too. Every musical note (i.e. every black and white key on the piano keyboard) has a pitch that corresponds to a very particular frequency, and we'll see later in this chapter what these are and how they relate to musical instruments, including drums.

In Chapter 2, we discussed the vibration modes of a drum, and we mentioned that the most powerful vibration of a drumhead is excited when we hit it in the centre, and that's when the drum's pitch is really quite obvious to hear. Hitting the drum in the middle causes the drumhead to vibrate up and down with a very powerful fundamental frequency (which we call F0 for shorthand) relating to the tension and other physical properties of the drumhead. This vibration causes a very small amount of air compression inside the drum, which translates onto the bottom drumhead, and ultimately the bottom drumhead moves up and down too at the same frequency as the top one, making a really powerful vibrating system with a sound we know as that of a popular drum.

3.1 Exploring the pitch range of a cylindrical drum

We know that changing the tension of the drumhead, by tightening or loosening the tension rods attached to a drum's lugs, changes the frequency and pitch of a drum's sound, so it's interesting to learn what the pitch range of a particular drum is. Imagine a drum has just had two new heads put on and they are both at the point where they are really loose, but just tightened so that they can vibrate when hit (i.e. all the slackness is taken out of the drumhead and it has no wrinkles in it), but each are as loose as possible without sounding broken. Since the pitch of the drum is related to the two drumheads vibrating together, if we tighten either drumhead, the pitch of the drum will go up. If we tighten both drumheads, it will go up more, so imagine tightening both to the point where the two drumheads really cannot be tightened any more, or they will snap and break; OK, that's now the highest pitch of the drum. In reality, these two low and high settings both sound a bit extreme, and there's certainly a sweeter sounding and slightly narrower range somewhere in the middle. But the pitch range for a cylindrical drum is quite big, sometimes covering a number of notes on the musical scale.[2] It's possible to experiment with this range to decide what pitch you want each of your drums tuned to, to give the best sound for your kit, the best sound for the style of music you are playing, and the best sound for your personal performance style and preference too. We know that there are lots of frequencies given off when the drum is hit, but the fundamental frequency which is heard when the drum is hit in the middle is the one that significantly defines the overall pitch of the drum, so you don't need to worry too much about overtones and the other frequencies when focusing on tuning the drum's overall pitch.

In an experiment to show an example of a drum's frequency and pitch range, a 13″ drum, using standard clear drumheads, was tuned from its slackest possible vibration frequency to virtually the highest tension it could go to without breaking. The results at each stage of tightening the batter and resonant drumheads are shown in Figure 3.1.

The data in Figure 3.1 is by no means scientifically gathered, since the adjustments at each stage are by hand and with simple human judgement; however, the tuning range of the drum can still be clearly evaluated. Looking first at the bottom left data point, we see that the drum's fundamental frequency has a lowest value of approximately 60 Hz. The fundamental frequency increases fairly linearly, i.e. with equal amounts, at each stage for adjustments to the batter or resonant tuning rods. The drumheads both approach a point close to breaking at a fundamental frequency of around 180 Hz. The drum used in this experiment didn't sound too great below about 80 Hz or above 150 Hz (which in itself is subjective!), but nevertheless, this gives a huge tuning range to explore and experiment with.

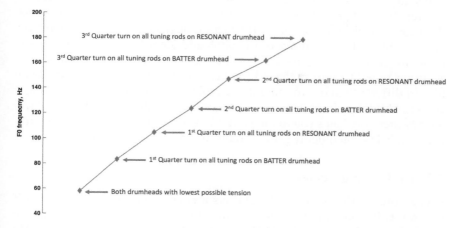

Figure 3.1 Tuning range of a 13″ tom with clear batter and resonant drumheads, with alternate tensioning on batter and resonant sides. All readings are taken with iDrumTune at the centre of the batter drumhead.

TRY FOR YOURSELF: PITCH RANGE OF DRUMS

Take one of your tom drums, place it on a snare stand, and loosen all of the tension rods until the batter drumhead is just able to vibrate when hit in the centre. Make sure there are no visible wrinkles in the drumhead. Do the same for the resonant drumhead. Hit the drum in the middle and listen to the sound - it should sound relaitvely low. Take a reading at the centre of the drumhead with iDrumTune (use the Pitch Tuning or Spectrum Analyzer mode) and make a note of the frequency. Now apply a quarter (90-degree) turn on each of the batter drumhead's tension rods with a tuning key. Take another reading and see how much higher the fundamental frequency of the drum is.

Now turn the drum over and tighten all the resonant side tension rods by a quarter turn. Revert back to the batter head and take another reading of the F0 at the batter head's centre. You'll see it's gone up again. Continue to make adjustments alternating between the batter and resonant drumheads, taking a reading of the batter head's fundamental each time.

Take each drumhead as high as you can safely go without getting too close to the breaking point. What is the usable pitch range of your drum? The drum should sound good at all stages, except perhaps for the extremes of the range, but obviously some frequencies will sound better with some styles of music than others. There are no rules here, but it's valuable to get an understanding of where you like to tune your own drums in terms of pitch – which pitch did you prefer the sound of as you moved through the tuning range? Of course, you can try this for your other toms too and also the snare with its snare wires off.

It's not uncommon for jazz drummers to use relatively high pitch tuning (nearer the tight end of the range), and rock drummers often go for something lower pitch and more powerful (nearer the slack end of the pitch range). It's valuable to take readings of the pitch and identify what frequency you prefer for each of your drums. Then you can make sure each of your drums is always tuned to your preferred frequency every time you play – or if you play in different bands, you might use different tunings for different music genres. Thinking about drums in terms of their pitch and fundamental frequency therefore helps you to understand the sonic differences between tunings and helps you quickly get the sound you need at any moment in time.

3.2 Musical frequencies

Every note on the piano keyboard has a corresponding frequency, so if you want, it's possible to tune a drum's fundamental pitch to be on the musical scale. For example, 98 Hz is note G2 on the musical scale, 110 Hz is A2, and C3 is at 130.8 Hz. Having some knowledge about these musical frequencies can help you to decide exactly what fundamental pitch you want each drum in your kit to be tuned to. Figure 3.2 shows every note on the piano within 50–400 Hz, giving the associated frequency value for each musical note.

There are a number of mathematical associations between musical pitch and frequency. One of the simplest to identify is that octave notes are always seen when the frequency doubles. Looking at Figure 3.2, we see that the C2 note has a frequency of 65.4 Hz, and this frequency doubles at the next C note (C3), which is at 130.8 Hz (65.4 × 2 = 130.8). The frequency doubles again at C4, which is 261.6 Hz, and you can see that all octaves of all notes occur when the frequency doubles.

It's also possible to use the frequency chart to help set up the intervals between the different drums in a drum kit. So, as you play a drum roll around the kit, you go up (or down) in frequencies at well-spaced intervals. If you tune to musical pitches, then you can start to develop really interesting phrases around the kit, and your fills will sound musical and stand out. Grammy winning producer/engineer Sylvia Massy emphasises the value of pitch tuning when referring to a particular project with the band Tool:

> Danny Carey from the band Tool is as important a composer in the band as the other players. For his drums I took a great deal of time, studying the song to be played and tuning the drums into the key of each song. The toms were tuned to make a chord, which helped to make his drums melodic and musical.[3]

We'll discuss this aspect of tuning in more detail in Chapter 9, which identifies different strategies for tuning musicality into the whole kit. Some

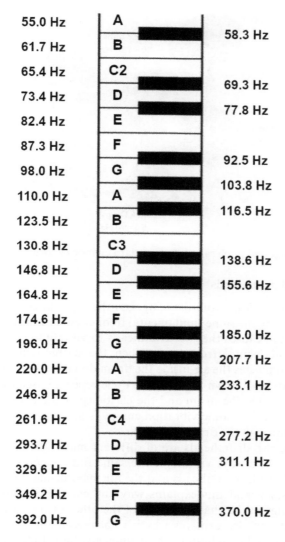

Figure 3.2 Frequencies corresponding to musical notes A1 to G4.

suggested frequency ranges for different drums in a standard type kit are shown in Figure 3.3, with rock sounds tending to be towards the lower end of each range and jazz sounds tending to be towards the higher end of each suggested range. The size of the drum makes a difference too, so a 16-inch diameter floor tom will generally have a lower tuning range than a 14-inch diameter floor tom.

Figure 3.3 Suggested pitch ranges for standard cylindrical drums.

3.3 Coupled drumheads

It may not be obvious from listening, but the top and bottom drumheads vibrate at exactly the same fundamental frequency. The mass of air trapped inside the drum causes the two drumheads to vibrate together at the most powerful fundamental frequency. When the batter drumhead vibrates downwards, it pushes the air inside the drum downwards, which pushes the resonant drumhead down too, and the cycle repeats based on the pressure disturbances inside the drum and the elasticity of the two drumheads, as shown in Figure 3.4. Some vibration energy is also passed on between the drumheads through the drum shell too.

Even if the two drumheads are of different materials, designs, or thicknesses, they will always vibrate at the same fundamental frequency. The drum, with both its drumheads and the air mass inside, forms a complete vibrating system, and this explains why any change on either drumhead causes a change in the characteristics of the other drumhead. The coupled drumhead theory also explains why it's not possible to accurately tune a two-headed drum by evaluating one drumhead alone. Indeed, if you place a drum on, for example, a cushion or a drum stool (to stop the resonant drumhead from vibrating), you'll immediately hear the huge difference in sound when hit, and if measuring the fundamental frequency with iDrumTune, you'll see that it changes dramatically too when dampening one drumhead. This is because you have now changed the mechanics of how the drum creates sound, in a way that is not related to how a drum creates sound when positioned on a drum kit during a performance. We don't play drums with one vibrating drumhead and one restrained, so it's therefore unwise to try to tune drums with one drumhead held fixed. This essential relationship between the drumheads is an integral fact of physics[4]; however, that doesn't mean the two drumheads sound exactly the same when hit with a drumstick.

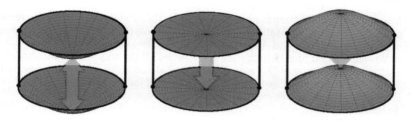

Figure 3.4 Coupled drumheads vibrating with a common fundamental frequency.

The two drumheads create a different sound when hit because the overtones on the top batter head can be completely different to those on the resonant head, even though the fundamental pitch of each drumhead is identical.

TRY FOR YOURSELF: COUPLED DRUMHEAD THEORY

We can easily prove the coupled drumhead theory by measuring the fundamental F0 frequency of a drum on both the batter and resonant drumheads. Take one of your toms and place on a snare stand, and take a reading of F0 with iDrumTune at the center of the batter drumhead, using the Pitch Tuning or Spectrum Analyzer mode. Now turn the drum over and take a reading of F0 from the centre of the resonant drumhead. You'll see they are exactly the same, indicating that the two drumheads vibrate together at a single fundamental frequency.

Make a small adjustment to all the tuning rods on either the batter drumhead or the resonant drumhead, and take readings of F0 on each drumhead again. The fundamental frequency will have changed, but it will still be the same when measured on both drumheads. Make some more changes and see that you just can't get them to give different F0 readings. There might be a small variation, by 1 or 2 Hz, but this is generally owing to differences in the drumhead designs or to small losses of energy as the mass of air moves inside the drum.

Now explore how the drum sound changes when you restrain or fully dampen one of the drumheads. Place the drum on your drum stool or a cushion, to dampen the resonant drumhead, and notice how big a difference it makes to the sound of the drum. The drum just can't vibrate and create sound properly, so it's impossible to evaluate accurately how it vibrates. Take a reading with iDrumTune with and without dampening the resonant drumhead, and you'll see the fundamental frequency changes by a significant amount – this evidences why it's not possible to tune drums accurately without both drumheads vibrating as they would in a performance situation. To ensure this is always the case when tuning, it's best to either position drums on a snare stand, or as they would be normally around the drum kit when you are performing.

Notes

1 Musical pitch perception is also influenced by the strength of harmonics related to the fundamental vibration frequency, as discussed in *Acoustics and Psychoacoustics* by Howard and Angus (2017), 5th Edition, Focal Press, p. 135.
2 A YouTube video of a 13-inch tom being tuned all the way from almost slack to the first good reading at 79 Hz, and then tuned all the way up to 138 Hz is available by iDrumTune at https://youtu.be/ITYRmvZ7ueI [accessed 01/08/2020].
3 Interview with producer/engineer Sylvia Massy conducted on 30/08/2020.
4 As discussed in Science of Percussion Instruments by Thomas D. Rossing, 2000, Volume 3, World Scientific Publishing Co., p28.

4 Lug tuning and clearing the drumhead

Lug tuning is sometimes called *equalising the drumhead* or *clearing the drumhead* and ensures that the tuning is even at every point around the perimeter of the drum. We often evaluate this at each of the lug positions around the drumhead and make adjustments with the tension rods that connect the hoop and drumhead to the lugs and the drum shell. Accurate lug tuning is important because, when completed, it allows the drumheads to vibrate evenly and smoothly. An uneven (or *uncleared* or *unequalised*) drumhead can cause unwanted frequencies to interact and cancel each other out, resulting in a pulsing, warbling, or beating effect on the drum sound. If you can hear a tone that comes in and out, sounding like a "wow-wow-wow" in the decay of the sound, then it is likely that at least one of the drumheads needs equalising.

The concept of lug tuning isn't exactly new – indeed, many drum tuning educators, for example Martin Ranscombe and Bob Gatzen, refer to lug tuning to achieve a drum sound that exhibits "a nice tone that decays with a smooth even note"[1] and to emphasize that "you are more able to hear a single pitch the more even your tuning is".[2] A number of scientific studies have also been conducted with respect to lug tuning, helping us to understand what exactly are the benefits of a drumhead that vibrates uniformly.[3,4] It's a simple concept, but one that is very difficult to implement accurately purely by listening to the drum. However, with a little musical acoustics knowledge, lug tuning becomes a powerful technique that can transform an uncharismatic sounding drum into one that sings!

4.1 Evaluating the first overtone of a drum

In the previous chapters, we discussed the differences between a drumhead's fundamental frequency F0 (excited at the centre and giving the BOOM sound) and the first overtone F1 (excited at the edge of the drum and giving the PING sound). The F1 overtone is the frequency to listen for when tuning at the lug positions and clearing the drumhead. This is because, while the F0 frequency is the same regardless of where you hit the drum, it is actually possible for the F1 frequency to be different at different locations around the edge of the drum. As mentioned, if each point around the edge of the drum has a slightly

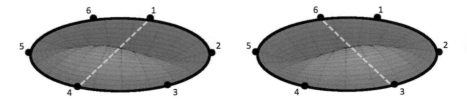

Figure 4.1 F1 overtone vibrations between different lug points on a drumhead.

different frequency, this causes a vibration clash on the drumhead which makes the sound of the drum warble or modulate as it moves. In acoustics, we call this condition *beating*, which occurs when two or more close, but not exact, frequencies are present at the same time. Beating in the drumhead means that the sound pulses loud and quiet in volume, so it is not smooth and you don't get a pure even tone from your drums when they are hit.

Let's look more closely at how this uneven condition is possible and the practical mechanics of how it can be corrected. We saw previously that the F1 vibration mode has a kind of see-saw profile, where the edges move up and down, but the middle stays stationary. In the condition where F1 overtones around the drumhead are not equal, the drumhead is attempting to vibrate at two very close but not exact frequencies, and, as a result, the frequency differences start to cancel each other out and clash. Figure 4.1 shows the overtone vibration in the diagonal axis between lugs 1 and 4 and also between lugs 3 and 6. If these two vibration frequencies, for example, are not identical, then the drumhead will vibrate unevenly. In many respects, the drumhead acts like an infinite number of tensioned strings positioned at diagonals across and through the centre of the drumhead. So you can hopefully imagine that a string between two opposite lug points can be tensioned differently to a string between two other opposite lug points. Our challenge with lug tuning is to make sure that all these points across the drumhead are tuned to all vibrate at the same common overtone frequency.

4.2 Beat frequencies

When two frequencies are very similar but not the same, as with the unevenly tuned drumhead example, they start off enhancing the overall sound. But gradually, as one vibrates faster than the other, they become more and more out of sync, to the point where they cancel each other out completely. They then start to gradually come back into sync with each other and the cycle repeats. This effect is shown in Figure 4.2, where a 110-Hz sine wave (top) and a 120-Hz sine wave (middle) are combined, with the beating effect seen in the resultant waveform after the two signals are summed or mixed together (bottom).

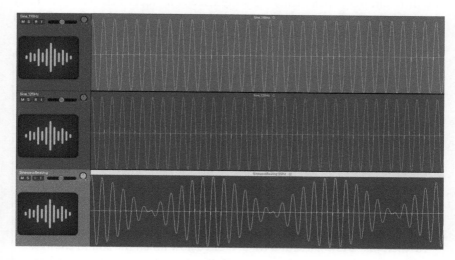

Figure 4.2 Two fairly similar sine waves (top and middle) combine to form a beating sine wave (bottom).

Figure 4.2 shows the beating effect in a DAW software package such as Logic Pro, but beating also happens in electronic circuits too; for example, small frequency inaccuracies from analogue circuitry can make tuning old analogue synthesisers very difficult. Beating also happens in acoustics, i.e. through the physics of vibration, as we have seen in the drumhead example. If you play the same note on different strings on a guitar, you'll hear beating if one is not perfectly in tune with the other. Beating also happens in air, as molecules collide against each other and cancel out their respective energy and motion, we'll prove this in a moment with a practical exercise.

Another characteristic of beating is the speed at which the beating, or *amplitude modulation* as it can also be called, occurs. The *beat frequency* is determined by the difference between the two sine wave frequencies, so the beat frequency of the example in Figure 4.2 is 10 Hz. This means that the volume of the mixed signal will modulate up and down exactly ten times per second. If the two signals had been closer, say 110 and 112 Hz, then the beat frequency would be 2 Hz, which would sound like a much slower increase and decrease in volume. Beating occurs because it's not possible for a signal to have two very close frequencies at the same time without those two frequencies interacting with each other. Equally our ears are not capable of distinguishing two frequencies that are very similar, which, as a side note, is what enables audio data compression algorithms like the MP3 to work. When hearing two frequencies that are very close together, our ears cannot distinguish them and instead we hear a single frequency with the beating effect applied. As the two frequencies become further and further apart, the beating starts to become more rapid, and eventually the fusion breaks, and all of a sudden the two different frequencies can be distinguished and identified. This is a really interesting aspect of acoustics and makes for a great practical example, and it's something to be very aware of when tuning drums too.

TRY FOR YOURSELF: SINE WAVE BEATING

Let's listen to some of these beat frequencies and connect the theory to our real-world hearing! In Logic Pro, set up two audio tracks and add the Logic Test Oscillator utility plug-in to each channel. Set both oscillators to the same frequency, say 110 Hz, which is an A note on the musical scale. You should be able to hear a pure sine wave with both oscillators active. Now increase the frequency of one oscillator to 111 Hz; you should immediately be able to hear the beating as the volume moves up and down every second. Increase the oscillator to 112 Hz and you'll hear the speed of the beating double. Now increase it further and further, just a few hertz at a time. Notice how much the sound changes with a beat frequency difference of 10 or 20 Hz compared to a beat frequency of 1 or 2 Hz. As you increase the frequency difference more and more, there comes a point where the beating is gone and now you can hear two pure frequencies. What difference was this frequency for you? You can also try this at different starting frequencies, such as 220 Hz, which is an octave higher.

We can also prove that beating happens acoustically, in real space, too, not just inside a computer. You'll need some loudspeakers for this, as it doesn't work as well with headphones. Hard pan one of the oscillator channels to the left and hard pan the other oscillator channel to the right, as shown in Figure 4.3, and you should be able to hear the beating occurring as before. Now press the mute button on one of the channels, and you'll hear a perfectly pure sine wave on one side of your playback system. Unmute and the beating will come back. Now try with the other side, pressing mute and unmute, and try with different frequencies too. It's a pretty cool acoustic effect happening right there in the acoustics of your room!

Figure 4.3 Logic Pro set-up for experimenting with sine wave beating.

4.3 The sound of a uniformly tuned drumhead

We've discussed what beating looks and sounds like on a pure sine wave, but it's also important to experience how this translates to a drumhead which is not cleared and has an uneven tuning around the perimeter. Figure 4.4 shows the recordings of drumstick hits at the edge of two different drumheads: one is evenly tuned (top) and one is unevenly tuned (bottom). It's very clear to see the beating effect in the audio waveform of the unevenly tuned drumhead, and it's very distinctive in the sound of the drum too.[5]

The beating effect is exactly the same effect as a tremolo guitar pedal or mixing plug-in. If your drum has two or more clashing frequencies, it can be very noticeable and ruins the smooth decay of the drum when hit. It's even possible to see the split beat frequencies identified in iDrumTune Pro, if you use the Spectrum Analyser mode. If the drumhead is unevenly tuned, then sometimes it is possible to see two F1 peaks very close together, which is a sure sign that the drumhead needs clearing or equalising. Figure 4.5 shows the difference in frequency spectra between an evenly tuned drumhead (Figure 4.5a) and one which is not evenly tuned (Figure 4.5b). The well-tuned drumhead on the left exhibits a strong F0 frequency at 102 Hz with a significant edge overtone at 170.5 Hz. In contrast, the drumhead on the right shows two very close frequency spikes in the F1 region (circled), indicating that the drumhead is attempting to vibrate at two close but different frequencies and needs to be equalised with lug tuning.

Unfortunately, because of potential beat frequencies and the fact that many different frequencies are excited on the drum at the same time, it can often be really hard to hear if different points on the drumhead are the same pitch or not. It can even be hard to hear if one is higher or lower than another, because our ears get drawn to different frequencies and our brain finds it hard to interpret all the information. In a recent listening test study, participants were asked to compare recorded drumhead sounds

Figure 4.4 Microphone recording of even (top) and uneven (bottom) tuned drumheads.

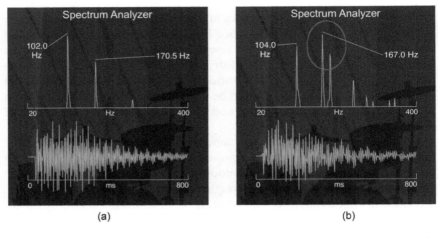

Figure 4.5 An evenly tuned drum gives (a) clear fundamental and overtone peaks on the frequency spectrum, whereas (b) an unevenly tuned drum can often exhibit two frequency peaks around the region of the first overtone frequency.

to identify if some had a higher, lower, or equal pitch, similar to the manual process of evaluating lug frequencies by listening to drumstick taps at various locations around the drumhead.[6] All participants were drummers, non-drummer musicians, and/or sound engineers with at least ten years of professional experience; yet, on average, they were only able to correctly identify the pitch differences 45.4% of the time. This is no major surprise given the complexities of drum sound; the harmonics (or lack of) from a vibrating drumhead make it very difficult to accurately identify pitch when listening specifically to the F1 overtone, since the fundamental frequency F0 has the potential to confuse the brain and influence our interpretation and decision making. We call this the *limits of acoustic perception* and the academic field of *psychoacoustics*, which relates to the subjective human perception of sound in comparison with the objective measurement of sound. In this context, the drum tuning listening test result validates the fact that there are only so many frequencies our brains can hear accurately at one moment in time – so don't feel inadequate if you find this aspect of drum tuning difficult!

4.4 Lug tuning with assistance

Given the limits of human hearing accuracy and the complexity of drum sound, it's valuable to have a more objective measure to help with lug tuning. The iDrumTune Lug Tuning feature provides just this, enabling a measurement of the drumhead's first overtone (F1) at all lug positions

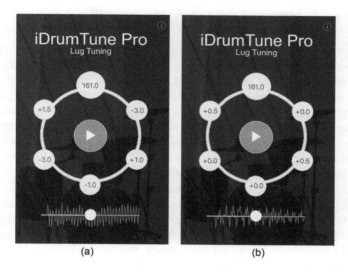

Figure 4.6 The iDrumTune Lug Tuning feature showing the software read-out for (a) an uneven tuned drumhead and (b) an even tuned drumhead.

around the drumhead. We generally compare all the lug positions with a first reference lug position, and then decide if any tuning alterations are required to achieve a more even frequency response around the drumhead. Figure 4.6 shows the iDrumTune Lug Tuning feature for a six-lug drum with both an unequalised drumhead (Figure 4.6a) and one which has been equalised to within 0.5-Hz difference (Figure 4.6b). In each case, the Lug Tuning feature gives the frequency of the first lug position, and then shows the deviation in hertz at all the other lugs with respect to the first lug position that was measured.

If the deviation from the first lug position is much greater than 1 Hz higher or lower at any other lug, then it's possible to make adjustments with a tuning key and gradually even out the drumhead. This can be quite a challenge, as the drumhead is very sensitive to even small changes in tension.

The benefit of using some electronic assistance, such as iDrumTune, for tuning around the lugs, is that it's programmed to ignore all of the other frequencies except the one that is most powerful, so it gives a reading and lets you know exactly which lug positions are lower or higher than the first one that you measure. You can then make some fine adjustments and measure again. If you get each point to within 1 Hz, then that is really pretty good, which is generally more accurate than most humans can discriminate. It's great to develop your hearing, and an app like iDrumTune can be used to check that you are making correct decisions, so you should test yourself from time to time by tuning by ear alone, and then using iDrumTune to check how close you are unassisted.

TRY FOR YOURSELF: LUG TUNING

Take one of your drums and identify the F0 fundamental (center) and F1 overtone (edge) frequencies with iDrumTune. Now go to the Lug Tuning mode, select the number of lugs that matches your drum, and take a measurement of F1 at each lug position. Readings displayed on the app in red (with a '+' prefix) are higher frequency than the first measurement, so these lug positions need loosening; blue readings (denoted with a '-' prefix) are lower frequency than the first lug position and so these need tightening. Make small adjustments and then take another round of readings of F1 around the edge of the drum.

You will find that you only need to tension each tuning rod very slightly up or down to even out the drumhead, and it is very easy to go too far and make a position that was too loose, suddenly too tight. Over time you will get a feel for how subtle you need to adjust the tuning accurately. If you get really stuck, then it can be valuable to untighten all the rods and then tighten them again with just your fingers (which gives a guaranteed even tension, albeit though slightly loose), before gradually increasing the tension with the tuning key and the Lug Tuning interface.

Now, purposely adjust some tension rods to make the drum unevenly tuned. Go around the edge tapping with the drumstick and see if you can hear any beating frequencies that go hand in hand with an unevenly tuned drumhead.

Something to be aware of is that it is essential that, when using a microphone to assist with lug tuning, you move the microphone to each lug point when taking readings around the edge of the drum. It's a simple fact of physics, but if you leave the microphone in one place when you hit the drum at different locations around the drum, the results are likely to be considerably less accurate. The reason is that the microphone acts as a very simple (non-contact) vibration sensor, and with lug tuning we are interested in how the vibration of the drumhead differs at very specific points around the drum. If the microphone is left in one place, then the sound recorded will be either purely a reading of that fixed location (every time, regardless of where you hit the drum) or, at best, an averaged reading of all the vibration characteristics of the whole drumhead. It's really important to get a very localised reading at each point for lug tuning, and that can only be done by moving the microphone to each lug position as you take measurements. It would be much easier to attach the phone to a single place on the drum and take lots of readings until you are happy, but, scientifically, that is not a valid approach and will lead you to thinking the drum is well tuned when actually it isn't. This issue with proximity between the sensor and the hit position also explains why drummers and musicians find it very difficult to clear the drumhead purely by listening, given that our ears are never really very close to the acoustic point of interest during lug tuning. It's equally important to ensure that you are not in some way loading or changing the characteristics of the drumhead when performing lug tuning. If you add any weight or damping to the drumhead when equalising the drumhead, then you are not evaluating the drumhead as it vibrates in reality or as it is played in the kit. So, in order to achieve accurate and reliable results, it's always important to perform this intricate tuning exercise with the drumhead vibrating freely and naturally.

You can also clear the resonant drumhead with the same lug tuning approach, but just make sure you tap very softly, as to not excite too many of the batter head frequencies and confuse the reading (and your ears!). If you have both the batter and resonant drumheads tuned to a uniform overtone frequency, then listen to the drum when you hit it cleanly in the centre and enjoy the smooth musical tone and controlled decay sound that it gives out.

Notes

1 Ranscombe, M. (2006), *How to Tune Drums, Rhythm Magazine*, Future Publishing, September, pp. 85–86.
2 Gatzen, B. (2006), *Drum Tuning (Sound and Design ... Simplified)*, Alfred Publishing Co., DVD.
3 Worland, R. (2010), Normal modes of a musical drumhead under non-uniform tension, *Journal of the Acoustic Society of America*, Vol. 127 (1), January, pp. 525–33.
4 Richardson, P. G. M. and Toulson, E. R. (2010), *Clearing the drumhead by acoustic analysis method*, Proceedings of the Institute on Acoustics, Reproduced Sound Conference, Cardiff, Wales, UK, 17–19 November, 2010.
5 Examples of stick hits on both even and unevenly tuned drumheads can be listened to in a YouTube video by iDrumTune at https://youtu.be/Wpxmam2D-MAo [accessed 01/08/2020].
6 *Evaluating the accuracy of musicians and sound engineers in performing a common drum tuning exercise* by Rob Toulson and Marques Hardin, Proceedings of the 149th Audio Engineering Society Convention, New York, October, 2020.

5 Tuning the resonant drumhead – what, why, and how?

Do you ever wonder why cylindrical drums have two drumheads, what exactly does the resonant head add to the acoustics of the drum, and how exactly does it influence the sound? Well, there are two main reasons. Firstly, a cylindrical drum with just one drumhead is very inefficient, energy wise, because as soon as the head is hit, all the energy created from the hit gets transferred into acoustic energy that leaves the drum and heads off into the space of the room or performance area. If we have a second drumhead, then a good proportion of that energy is reflected back inside the drum, and the two drumheads vibrate together, holding the energy within the air that is trapped inside the drum. This makes drums resonate for longer and the drumheads interact more with the drum shell, allowing more powerful, characterful, and louder sounds to be emitted by the conservation of energy within the drum. The second reason why cylindrical drums have two drumheads is owing to the sonic influence the resonant drumhead has on the overall drum sound and the vibration characteristics of the drum as a whole, which we explore in detail in this chapter.

5.1 Harmonics and in-harmonic overtones

Before explaining the resonant drumhead's influence on drum sound, let's just consider the vibration profile of musical strings (as in guitar) and bars (e.g. metal glockenspiel bars or wooden xylophone bars). It's an incredible phenomenon of physics, but strings and bars vibrate with perfect *harmonic* overtones. This means that the main fundamental frequency of a string or bar is joined by many other frequencies that are harmonically related, which results in a beautiful rich tone that is much more "musical" than a single frequency all on its own. The additional overtones are, by a chance of physics, at perfect multiples of the fundamental frequency, so a string tuned to A at 110 Hz also vibrates at harmonics of 220, 330, 440 Hz, and so on (as shown in Figure 5.1), so does a bar tuned to the same frequency. This acoustics fact is what makes string and tuned percussion instruments so musical sounding. In fact, the same principle applies to the vibration frequencies on woodwind and brass instruments too!

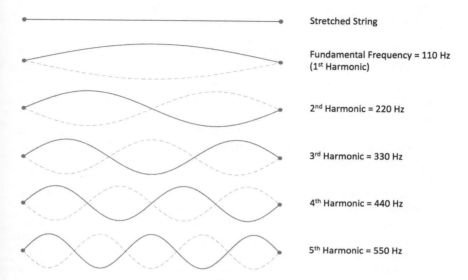

Figure 5.1 Harmonics of a vibrating string with a fundamental frequency of 110 Hz.

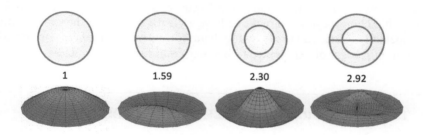

Figure 5.2 Frequency modes and associated frequency ratios of a single ideal drumhead.

Unfortunately, single tensioned drumheads, or *circular membranes* as we call them in acoustics, do not vibrate with perfect integer harmonics; they vibrate with what we call *inharmonic* overtones that are not related mathematically to the fundamental pitch. Hence, drums don't generally sound as musical as string or woodwind instruments. Actually, much research has been conducted on this subject, and we see that single drumheads vibrate at seemingly random spaced overtones of the fundamental frequency, none of which are musically related to each other.[1] Where the tensioned string has harmonic overtones at nice integer ratios of 2, 3, 4, 5, etc., the taut drumhead has its first frequency ratios at 1.59, 2.30, and 2.92.

In Figure 5.2, we see that each frequency ratio is fixed for a single drumhead, so it cannot be tuned to have overtones at harmonic or musical intervals. For example, an ideal drumhead (i.e. a mathematically modelled drumhead) with a fundamental frequency of 100 Hz will have overtone frequencies at 159, 230, and 292 Hz, none of which are musically or harmonically related to the 100 Hz fundamental. In reality, drumheads don't follow the exact frequency ratios of the ideal drumhead, because they all have different mass, thickness, and elasticity. The main concept and principle holds true however, meaning that there is very little that can be done to change the sound profile of a single drumhead.

But, with a second drumhead on the drum, it is all of a sudden possible to manipulate the relationship of frequencies and tune so that some of the frequencies, particularly the most powerful ones, do fall into musically related values. As a result, two-headed drums can be tuned to resonate with a much more musical tone, which gives the drums a more interesting, warm, controlled, and musical sound. It is not possible to make all of the drumhead frequencies harmonically related, so drums will never sound as musical in terms of pitch and resonance as guitars or clarinets, but it is a small improvement that is worth making. As a result, two-headed drums have become pretty standard in pop and rock music, and many drummers have learnt to tune drums until they sound good on the aforementioned principles, without really having an urgent need to understand the underlying science. Understanding the science is extremely valuable for any musician though, as it helps you to understand what characteristics to listen out for and what changes you can make to have a positive effect on the sound – this is especially true if you are learning drum tuning and you don't want to wait to build up years of experience to become confident and proficient.

TRY FOR YOURSELF: SINGLE DRUMHEADS

To hear the difference and value of the second drumhead, just take off a resonant head sometime when tuning, and you'll immediately hear the value that the second head brings, in terms of both power and tonality. If you have two similar sized drums, it can be very interesting to stand two side by side, one with just a single drumhead and one with both drumheads. It's even possible to use iDrumTune to tune both to the same F0 frequency, and then you can make a fair sonic comparison of the sound differences between the two. As you hit one and then the other, think about all the qualities in the drum sound. The two-headed drum will have a stronger, fuller sound with more interesting overtones. The single drumhead set-up is still useful, and many drummers play with just one drumhead – if it sounds right, it is right! But there is much less opportunity to control and manipulate the sound with tuning when using single drumheads.

An interesting side note at this point is that tympani and tabla instruments, for example, only use one drumhead, but are designed to have a very specific shape and construction that enhances the musical pitch of the drum. The detailed design of the drum, using a closed bowl rather than an open cylinder, is another method for improving the harmonic nature of the overtones of a drum.

5.2 Musical intervals

For cylindrical drums, it's valuable to know what the acoustic relationship between the two drumheads is and ensure that they are sensibly tuned relative to each other. A convenient fact for tuning the resonant drumhead is that, because the resonant head has an influence on the response of the batter head, it's not actually necessary to measure the frequencies of the resonant head itself to get an understanding of how it is tuned. It is therefore possible and interesting to assess the batter head frequencies and see how they change when tightening or loosening the resonant head. The resonant drumhead therefore controls the relationship between the batter head's F0 and F1 frequencies, and this is more than enough information to develop a tuning strategy for drums.

Let's look at this in practice. If we take a reading of the batter head in the centre, let's say it has a F0 reading of 100 Hz. Now, taking a reading of the batter head's edge F1 frequency, let's say it gives a reading of 160 Hz. A quick bit of maths tells us that F1 is 1.6 times greater than F0. Ok, so how is that information useful? Well, when we discussed string vibration frequencies, we noticed that they tend to vibrate with exact harmonics of the fundamental, i.e. two, three, or four times the fundamental; perfect harmonics are not really possible to achieve with drums, but charismatic *musical intervals* are!

Musical intervals define the relationships between frequencies in a musical scale. Looking at this on the piano keyboard, we see from C to C there are 12 semitones (i.e. 12 piano keys), but a major scale has only eight notes, those being C-D-E-F-G-A-B-C for the scale of C major. Figure 5.3 shows the

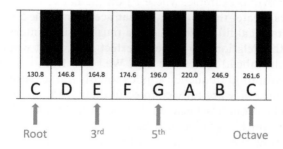

Figure 5.3 C major scale and associated frequencies in hertz.

Table 5.1 Major scale musical intervals and associated
frequency ratios

Major Musical Interval	Frequency Ratio
First (root or fundamental)	1.00
Second	1.12
Third	1.26
Fourth	1.33
Fifth	1.50
Sixth	1.67
Seventh	1.89
Eighth (octave)	2.00

frequencies of each note in the C major scale, which are all musically and mathematically related to the root or first note in the scale. For example, we call the third note in the major scale the major third and the fifth is the fifth note in the scale, which is G in the scale of C major.

By looking at the frequency ratios (i.e. mathematically dividing one frequency by the other), we can see the multipliers for each note in the scale. For example, we see that the fifth of C3 (130.8 Hz) is G3 at 196.0 Hz, and some simple math shows that 196.0/130.8 gives a frequency ratio of 1.50. Similarly, the frequency ratio of the major third is 164.8/130.8 = 1.26, and the frequency ratio of an octave is exactly 2. These frequency ratios are what define musical intervals of a particular scale, and it's possible to tune a two-headed cylindrical drum with these frequency ratios in mind, in order to achieve a more musical, rich, and well-balanced sound to the drum. The full list of major musical intervals and their associated frequency ratios is given in Table 5.1.

5.3 Controlling overtones and intervals with the resonant drumhead

So how does someone go about tuning the resonant drumhead sensibly and repeatably, taking the musical frequency ratios into account? Let's look back to our previous example, where F1 on the batter head is measured as being 1.6 times greater than F0. Ok, we can see from Table 5.1 that 1.6 is a bit in between a musical fifth interval and a musical sixth interval. So, assuming we want this relationship to be a perfect fifth, what would we do with the resonant drumhead to achieve this? We know from Chapter 2 that the fundamental vibration frequency applies equally to the batter and resonant drumheads, since they are fully coupled in their vibration mode. However, there is much less coupling between the two drumheads with respect to the F1 edge mode, which is significantly more localised to the batter drumhead. So, by tightening one drumhead and loosening the other, it's possible to make the batter head's F0 and F1 frequencies come closer together or move further apart, which is a powerful concept in controlling the overall tone of a drum. The relationship between the two batter head frequencies is referred to as the *resonant tuning factor* or *RTF* for short, since the resonant head tuning allows this value to be made greater or smaller.[2]

In practice, to manipulate the relationship between the batter head's F0 and F1 frequencies (i.e. to change the RTF value), it's possible to show that:

- To increase RTF: increase batter head tension and loosen resonant head tension
- To decrease RTF: loosen batter head tension and increase resonant head tension

---

TRY FOR YOURSELF: CONTROLLING RTF

It's possible to use iDrumTune's Resonant Head Tuning feature to identify and control the RTF value of a drum. First, take a tom drum with two heads, place on a snare stand, and make sure both are fairly well equalised with Lug Tuning. Then in iDrumTune's Resonant Head Tuning mode, take a reading of the batter head at the centre (F0) and then on the batter head at the edge (F1). Make a note of the RTF value given by iDrumTune. You should now be able to change the RTF value by tightening and loosening the batter and resonant drumheads as described above.

First, tighten all the batter head tuning rods by a small amount, then turn the drum over and loosen all the resonant head rods by a small amount. Go back to the batter head and take another RTF reading. You should see that the RTF value has increased. If you repeat this process, you'll see the RTF value go even higher.

Now you can try the opposite approach – loosen the batter rods and tighten the resonant rods, and see the RTF value come down.

What are the highest and lowest RTF values you can achieve with your drum? Did you have any preference of which RTF value gave the drum a best or preferred sound to your ears?

Figure 5.4 shows the iDrumTune Resonant Head Tuning feature. First, the app informs you to hit and read the batter drumhead in the centre (Figure 5.4a), then take a reading at the batter head's edge (Figure 5.4b). The RTF value is then calculated and gives advice on how to adjust the batter and resonant drumheads in order to achieve an RTF value of around 1.5 (Figure 5.4c).

A good starting recommendation is to aim for an RTF of 1.5 for two reasons. Firstly, this represents a perfect musical fifth, which is a good place to begin with when deciding how you want your drums to sound. The second reason is related to a rather interesting psychoacoustics phenomenon which

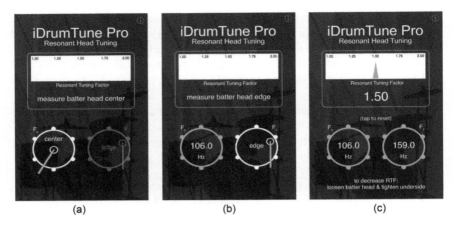

(a)   (b)   (c)

Figure 5.4 iDrumTune's Resonant Tuning feature used to calculate RTF values for a drum.

we might call a *phantom sub-harmonic* or a *missing fundamental*. If we have a frequency and another frequency at 1.5 times the first, then our brain anticipates that these are related by an even lower frequency at 0.5 times the first frequency, which gives the sense of a low-frequency power to the sound.[3] This lower frequency isn't actually there; it's an illusion on our hearing, but sounds great all the same and makes us perceive the drums as having a bit of extra low-frequency power. It's also the same clever technique used by loudspeaker manufacturers to make very small speakers sound like they have good low-end bass response!

In reality, it doesn't actually matter what RTF value you tune to, but it makes sense to have this somewhere between 1.3 and 1.8 and ideally to one of the musical benchmarks in that range. Higher than that starts to get towards an octave and miss out on the musical richness of an interval overtone, and lower than that starts to get too close to the main fundamental. It also makes sense to have all drums use a similar RTF in the kit, to ensure that the tone structure of each drum is similar and consistent. A beneficial aspect of using the RTF value is that when changing drumheads, it's possible to immediately get back to the same sound for a set of drums as before changing heads. So if you use the same drumheads, the same fundamental pitch, and the same RTF value, then your drums will always sound the same, day-in and day-out.

TRY FOR YOURSELF: CONSISTENT RTF FOR ALL DRUMS

It makes sense to have all the drums in your kit use a similar RTF value; this way you'll have a consistent sonic character to all the drums in your kit. While each drum will have a different fundamental pitch, they will all sound like they belong together if they have the same tuning relationship between the batter and resonant drumheads. Indeed, if you have one drum in your kit which just doesn't sound right, or you struggle to get a perfect sound out of, there's a good chance that it has a very different RTF value to the other drums in your kit, and that's why it sounds out of place. Try to tune your rack and floor toms to have a similar RTF value. You may like to use the RTF = 1.5 recommendation (musical fifth) or maybe you prefer the musical sixth (RTF = 1.67), or maybe somewhere else in the range between 1.3 and 1.8. Regardless, experiment with tuning all drums to the same RTF and see how they sound as a unit, and you should find the kit starts to feel much more like a single instrument if all drums use the same type of drumheads and are tuned to a similar RTF value.

Notes

1 Modal ratios of an ideal (mathematically modelled) drumhead given by Thomas D. Rossing in *Science of Percussion Instruments*, 2000, Volume 3, World Scientific Publishing Co., p. 6.
2 The RTF and its use for drum tuning were first described in the Audio Engineering Society 2017 Convention presentation and paper titled *The Resonant Tuning Factor: A New Measure for Quantifying the Setup and Tuning of Cylindrical Drums* by Rob Toulson (New York, October 2017).
3 As described by Thomas D. Rossing in *Science of Percussion Instruments*, 2000, Volume 3, World Scientific Publishing Co., p. 9.

6 A holistic approach to drum tuning

The previous chapters have covered the most essential acoustics theories relating to drumhead vibration, so now it's possible to bring these all together in one place to consider a holistic, complete, and informed approach to drum tuning. It's true that drum tuning can be quite daunting for a new drummer. It takes a number of years for most drummers to develop the required understanding and hearing skills in order to tune drums unassisted. Many drummers, even those with lots of professional experience, find drum tuning a challenge and find it difficult to understand what physical changes in the set-up or tuning will achieve the sound they are looking for. Don't be ashamed of this; cylindrical drums are one of the most challenging instruments to tune and optimise the sound of. It takes a very thoughtful drummer or sound engineer to manage and control the sound of drums, but having some knowledge of the fundamental pitch, the value of uniform tuning, and the relationship between the two drumheads makes it possible for anyone to tune drums with control, accuracy, and consistency.

The holistic approach to tuning has four steps – three of which we have already discussed, and a fourth related to the use and addition of damping to control the sustain and decay times of different frequencies from the drum and drumhead. The four holistic drum tuning steps are therefore as follows:

1   Tuning the fundamental pitch of each drum in the drum kit
2   Equalising the drumheads to give a clear and smooth tone
3   Relative tuning of the batter and resonant heads
4   Controlling the decay and damping of the drumheads

There is really no need to overcomplicate things with tuning, especially once you understand the core concepts. With some knowledge of drum acoustics, it's very possible to keeping things simple, quick, and easy![1,2]

6.1 Simplicity wins

Drum tuning is quite difficult initially, but, over time, what seemed difficult at first will become more obvious and easy. It's a reflective learning process that all drummers go through in order to find their own personal sound and style. Drummers and sound engineers also develop their hearing skills with respect to percussion sounds over time, and understanding a little of the science behind the sound makes it possible to develop these skills more quickly and to eventually be able to tune without any significant help from tuning aids.

The best way to get to a good starting point is very simple. If you are tuning a drum kit for the first time, or you are putting a full new set of drumheads on your kit, it is best to start thinking about tuning when you first put the tuning rods into the lugs on the drum. So, with all the tuning rods removed, place the drumhead on the drum shell, place the hoop over the edge of the drumhead, and drop a tuning rod into each tuning lug. Be careful to ensure that the drumhead sits flat and evenly on the bearing edges of the drum shell. Now with just your fingers only, tighten each tuning rod until you cannot tighten it any more. It's often best to tighten two positions opposite each other at the same time, one with each hand, or tighten each rod a little and then go around the drum for a second and a third time if necessary. Once you have tightened the rods by hand and can tighten them no more with your fingers, you can now be sure that you have a fairly even tension applied to the drumhead at each lug position. You can use the same approach for both the batter and resonant drumheads.

Now, with the rods just *finger-tight* the drumheads will usually still be too slack to vibrate properly (you will, no doubt, see some wrinkles in the drumhead), so we need to tighten each tuning rod a little more before we can get the first reasonable sound from the drum. With a drum key, give each tuning rod a quarter or half turn of the key – i.e. 90 or 180 degrees. The choice of whether to give a quarter or half turn depends on the type of screw thread on your tuning rods, so it's wise to try a quarter turn first, and if the drumhead is still slack and showing wrinkles, then give another quarter turn to each tuning rod. Do this quarter or half turn on both the batter and resonant sides of the drum. Don't go around the drum, jump opposite in a star-form every time to ensure that all sides of the drum are kept similar at all times. Some star-form diagrams for drums with different numbers of lugs are shown in Figure 6.1, though it's ok to follow slightly different star-form paths than those examples given. Following the star-form approach is quite important when you are making significant changes to the drumhead tension. As we discussed in previous chapters, the drumhead vibrates somewhat like a number of infinite strings all connected to the circumference through the centre of the drum. Each point at the drum's edge therefore has a tension relationship with the opposite position on the other side of the drum. Imagine when tuning the drum you are actually tensioning a number of guitar strings connected between the lug points on the drum, it's important to tighten each of these "strings" together at each end and then move on to the next. Of course, the drumhead is much more complex than a few connected strings, because there are also tension relationships moving out in all directions from any one lug position, but it makes good practical sense

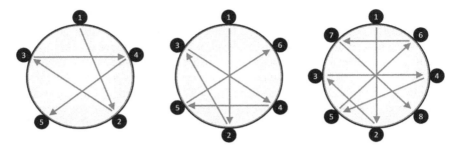

Figure 6.1 Example star-form tuning order for five, six, and eight lug drums.

to tighten each set of opposite tuning rods before moving on to the next opposite lug points. This approach also ensures that each side of the drum maintains a similar tension and the drumhead tightens in the most uniform way possible, meaning that there will be less drastic action needed when implementing lug tuning later on.

TRY FOR YOURSELF: FINGER-TIGHT PLUS A QUARTER TURN

Take one of your tom drums and loosen every single tension rod until it is fully slack on both the batter and resonant sides. Now seat the batter drumhead flat on the drum shell's bearing edges, and apply the finger-tight technique to each tuning rod. Repeat this for the resonant drumhead too. Then apply a quarter turn in the star-form approach to both drumheads.

Hit the drum in the middle of the batter head, and it should sound pretty good – drum tuning can be that simple! If the drum still sounds very low pitch and has a dull thud kind of sound, then perhaps the drumhead still isn't tight enough to vibrate properly when hit. In this case, give each batter-side tuning rod another quarter turn. How does it sound? If you want to tune it up a little higher, tighten the resonant tuning rods by another quarter turn each too. With some small adjustments on the batter and resonant sides, applying the same small turn to each tuning rod, you should be able to get the drum sounding pretty good.

This approach gives equal attention to every tuning rod and both drumheads, so you can be sure that the lug tuning and resonant head tuning aspects are both in a fairly sensible place. If you are tuning your first ever drum or you want drum tuning to stay really simple, then this is a great method to get drums sounding good with very little time or effort!

6.2 Setting the fundamental pitch

With the drumheads in place and tensioned to the first usable sound, the next thing to do is to find a good overall pitch for the drum – which basically means tuning the F0 fundamental frequency, heard when we hit the drum in the centre of the batter head. There is no right or wrong frequency; rock drummers tend to tune low and jazz drummers tend to tune high, so this is where a bit of experimentation comes in. See how high you can take the drum by making similar adjustments to each lug position as you did before; again this keeps the tension guaranteed to be fairly consistent at each point if you turn each tuning rod by the same amount. You'll probably want to be somewhere in the middle of the range between too slack and too tight, but this differs for every drum and genre and style, etc., but sounds good at many different frequencies.

The tuning ranges for standard-sized toms and snares differ a little depending on what drumheads are used. Thicker and heavier drumheads work well at low frequencies, but don't tend to be able to tune quite as high as thinner clear drumheads, which is something discussed in much more detail in Chapter 7. For now, Figure 6.2 gives a good overview of the kind of frequency ranges you can try for different sized drums, and it is worth experimenting with the range of your own drums to see which areas of the frequency chart best suit your own set-up and your own sonic preference.

Note that for snare drums, it's best to perform the initial tuning with the snare wires off or disengaged, so that you can make an accurate judgement of the pitch and tuning of the drum.

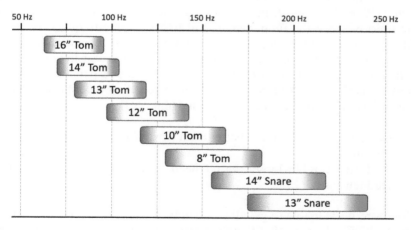

Figure 6.2 Frequency chart showing indicative tuning range for a number of standard-sized toms and snare drums.

6.3 Implementing lug tuning and resonant head tuning

Once you have control of the drum's overall pitch, it's time to check how evenly it is tuned around the lugs on the batter head. Using the iDrumTune app in the Lug Tuning mode, or purely by listening, hit the drumhead gently at the edge near a lug position and then listen or take a reading at each lug point around the drum. Even though a very consistent approach had been taken to adjusting each tuning rod in the prior steps, no doubt you'll identify that some lug frequencies are a little different to others. So you can nudge some positions tighter or looser with the drum key to get a more even result. This can take a few attempts going around the drum, but if you are within 1–2 Hz, then this is really pretty good. With lug tuning, since we are actively trying to change the balance of tuning to get a more even vibration profile, it's not as important to follow the star-form tuning method as before; if you are only making very slight adjustments to one or two tuning rods, then there is no issue with the drumhead becoming more uneven owing to the order of which tuning rods you adjust.

Of course, nudging the lug position frequencies might change the over-all pitch of the drum, so now you need to double-check that and bring everything up or down again if it's gone away from your preferred sound. The thing about drum tuning is that "everything affects everything", so there is always a little backwards and forwards involved in getting a great tuning, setting the pitch, then checking the lugs, then the pitch again, but once you have the hang of it, you'll do this all quite quickly and instinctively.

Now that you have a drum with good pitch and equal lug tuning, it should sound pretty great – and we haven't even touched the underside yet! In fact, if you want to stop there, then do so, the resonant head doesn't need tuning to the same level of accuracy as the batter head. However, if you want to flip the drum over and check the lug tunings of the resonant head, then that doesn't hurt.

As mentioned in Chapter 5, it's also valuable to check the resonant tuning factor (RTF) for each drum in your kit. This is conducted by evaluating the ratio between the fundamental and first overtone frequencies. Recall that a ratio of 1.5 represents a fifth and a ratio of 1.67 represents a sixth on the major musical scale. We recommend having all drums in your kit with a similar RTF, and the major fifth value of 1.5 sounds particularly good for most musical styles. Additionally, it's worth recalling the adjustments you can make to modify the RTF value for a drum:

- To increase RTF: increase batter head tension and loosen resonant head tension.
- To decrease RTF: loosen batter head tension and increase resonant head tension.

Moreover, it's valuable to get all the drums in your kit to have the same or similar RTF values, so that there is a musical congruence and a consistent character to your drum sound as you play around the kit.

TRY FOR YOURSELF: FULL-KIT TUNING

You should now be able to tune a full drum kit to sound powerful, smooth, and consistent. Ignoring the kick drum for a moment, start with getting every other drum to be finger-tight plus a quarter turn on both drumheads. Adjust the tension of each drumhead up to roughly the pitch you think sounds good for your kit and musical style. Then check and adjust each drumhead for a uniform and equal response with lug tuning. Check the RTF value on each drum and make some small adjustments to ensure all drums have a similar RTF value. You might need to go around the loop a couple of times with each drum, because every adjustment affects all aspects of the drum sound, but soon enough you'll get to a point where the drums are sounding powerful, smooth, and consistent. If you've used the right drumheads for your style, then this will be not far off a perfect drum sound!

Make a note of the F0 and F1 frequencies on each of your drums, and then you can experiment with trying some different tunings too. Maybe take all the drums up a little in pitch and see if you enjoy the extra musicality of the instrument. Or you might like to see how low the whole kit can go without sounding too dull. It's good to explore the range and understand when or why you might use different tunings in different circumstances. If you made a note of some of these frequencies, such as the first one, then you should be able to dial in different tunings very quickly after a little practice.

6.4 Damping and decay times

The final sonic aspect with tuning a full drum kit is to maintain control over each drum's sustain and decay time. As will be discussed in Chapter 7, sustain and decay times of the drum are best controlled by choosing the right drumheads for your preferred musical genre or drum sound. Many drumheads have advanced damping and sustain control systems built into their construction, meaning that, if you have the right drumheads on your kit, you really should have no need to use any additional form of damping added to the drumheads. Indeed drum tuning educator Martin Ranscombe emphasises this point in his well-cited *Rhythm Magazine* column from 2006:

> If your drums are well tuned and with the correct head choices, there really should be no need for additional dampening... no o-rings and absolutely no pillows, towels or other such stuff.[3]

Whilst this is very true, it is also the case that sometimes you just don't have the opportunity to change drumheads in order to resolve a decay issue.

Maybe you are playing for a different band as a one-off or playing a venue's house kit which is not to your taste; maybe the venue has a big boomy space that causes your drums to ring out for an eternity; or maybe you can't afford new drumheads or just don't have time to change to new drumheads at that particular moment. With this in mind, it's worth understanding what the sonic characteristics are of a drum's sustain and decay, and what you might be able to do to improve a drum that rings out too long after it is hit.

First let's just have a look at what we mean by sustain and decay. They kind of mean the opposite thing: either the drum sound sustains for a long period of time or it decays back to silence quite quickly after it has been hit. Some instruments such as a trumpet or a synthesiser are able to sustain for long periods based on how long the note is held, yet drums and most percussion instruments tend to have a sustain period based on the vibration characteristics of the instrument; essentially as soon as the drumstick stops being in contact with the drumhead, its energy and volume will gradually decay, either slowly or fast, back to silence.

Drum sounds which have a slow decay time, i.e. a relatively a long sustained sound after the initial hit, can sound great. The longer note allows the human ear time to better identify with the pitch and musicality of the sound. This can sound just right for jazz and musical styles which want to carry a very clear musical pitch to the percussion performance or for slow tempo ballads with emotive drum fills around the toms. However, long decay profiles can cause problems for drum sound, and are something which many drummers, drum makers, and drumhead manufacturers have long battled with. If the drum decay is too slow in comparison with the tempo of the song, it can bleed into adjacent notes, causing the drums to sound cluttered and less precise. Furthermore, if a long decay sound is not perfectly in key with the song, musical clashes can be heard between the drums and other instruments in the performance. Possibly the most challenging aspect of controlling decay times is the fact that the drum's frequencies rarely all decay at the same rate, so the character of the drum sound changes as it decays. Usually this is an issue because, for some drums and drumheads, the F0 fundamental decays quite quickly, whereas the F1 overtone rings out for longer. The result is a sound that loses its deep tone quite quickly and leaves a rather thin and high-frequency ring, so the drum's sound appears to be less powerful and without a rich character. We can see the effect of a ringing overtone in a three-dimensional frequency chart, often called a *spectrogram* or *waterfall plot*. The waterfall plot shows the same information as iDrumTune's Spectrum Analyzer feature, except with an additional time axis, so we can see what frequencies are present in the sound, and we can also see how those frequencies change over time. From the waterfall plots in Figure 6.3, we can see the difference between a drumhead that has a very strong overtone that decays along with the fundamental (Figure 6.3a) and a drumhead that has a much less powerful overtone that decays much more quickly in comparison with the fundamental (Figure 6.3b). In many music genres, such as rock, pop, and metal, the more controlled overtone is preferred, particularly on the snare drum which is often tuned to have a very

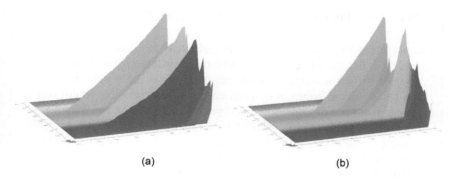

(a)　　　　　　　　　　　　　　(b)

Figure 6.3 Spectrogram comparison between drum sounds that have (a) significant overtone frequencies and (b) controlled overtone frequencies that decay more quickly.

sharp and rapid attack when the drumstick hits and rapid decay to the sound too. In the recording studio, microphones that are positioned towards the edge of the drum have a habit of accentuating the overtone unnaturally too.

There are many methods for controlling the decay of the drumhead, though none better than those which are built into the drumhead themselves. It is possible to use *damper rings* (or *o-rings*) which sit on the drumhead and add extra damping to the outer edge. This causes the overtone to be controlled yet allowing the fundamental to still retain its power. Heavier and wider damper rings enable the F0 frequency to be damped also somewhat if the decay time of the whole drum is a problem. A number of other solutions exist, which can have a more detrimental effect on the drum sound. Using damper gel, tape, or any other objects attached to the drumhead in a non-symmetrical way will have the effect of damping the drumhead; unfortunately, this is often at the expense of the tone of the drum. Adding a small amount of weight to one side of the drumhead causes the drumhead to vibrate in an unequal manner, and so, whilst helping to control the decay, undoes all the good work that has been done in lug tuning and equalising the drumhead, though this can work as a creative solution in some circumstances. Damping is a significant aspect of drum sound, so it's useful to understand and experience how drum sounds change in different performance spaces and with different drum set-ups. The type of metal or wooden hardware on a drum and even the type of rubber feet on a floor tom can have a subtle but influential effect on the decay time of the drum sound, so it's worth being prepared for changing heads (if time permitting) or adding some additional damping if the performance scenario calls for it. You'll also notice that some rooms accentuate the tone and decay times of some drums depending on the size and construction materials in the room. So you may find that some physical spaces require you to be more liberal with applying additional damping, in order to keep the sound controlled and to your preference.

TRY FOR YOURSELF: DAMPING AND DECAY

Experiment with different damping and decay methods, for occasions where it is not possible to change to a more suitable set of drumheads. A set of different weighted and size o-rings can be a valuable addition to any drum room or cymbal bag. It is also possible to cut some o-rings from old drumheads too if you need to improvise. You'll find that higher pitch drums tend to ring out for longer, so it's possible to use external damping to create a high pitch, yet heavily damped sound, which has a unique sonic character.

Different RTF factors will sometimes give different decay characteristics to the drum too, so listening to how the decay changes when you are tuning the resonant drumhead can be a valuable exercise.

It's worth comparing the sound of o-rings to the damping achieved by adding tape or gel to a drumhead. Often the tape or damper gel can provide a more substantial decay effect, but results in a return of the beating sound which was eradicated during lug tuning. It's therefore worth understanding the compromises associated with different solutions and the subtle sonic nuances of each type of external damping system.

You might also try more creative damping approaches, such as adding a fairly heavy weight or a cloth over the drumhead; these approaches break all the acoustics "rules" of tuning, but can sound really great in certain situations.

Notes

1 See *Drum Tuning 101: Back to Basics* on the iDrumTune YouTube channel, available at https://youtu.be/e_R3yLLihcE [accessed 01/08/2020].
2 Rob Brown published a number of YouTube videos on drum tuning, always promoting the idea of keeping things simple, quick, and easy. An example of Rob Brown's quick and easy tuning approach is available at https://youtu.be/lLEjrq_TFRg [accessed 01/08/2020].
3 Martin Ranscombe in his 2006 education series titled *How to Tune Drums*, published by *Rhythm Magazine*, Future Media, Summer 2006, pp. 87–88.

7 The wonderful world of drumheads

Being a drummer is both a blessing and a curse! The great thing about drums is that there are far more ways to personalise a drum kit than there are for any other musical instrument – the number of permutations of drum types, sizes, depths, materials, and configurations is endless. But the counter-side to this is that it takes a huge journey of learning, testing, and experiencing different set-ups in order to find your personalised "perfect" sound. No single aspect of the drum sound can be personalised more so than by the drummer's choice of drumheads used on the kit. Great drumheads and knowledgeable tuning can turn even a cheap battered drum kit into a hard-core rock-and-roll machine! Equally, with a well-crafted drum kit, it's possible to create impactful and musical sounds by choosing the right drumheads for your style and tuning them effectively. It's very clear that the type and size of drums you choose to use have a big influence on the sound of your

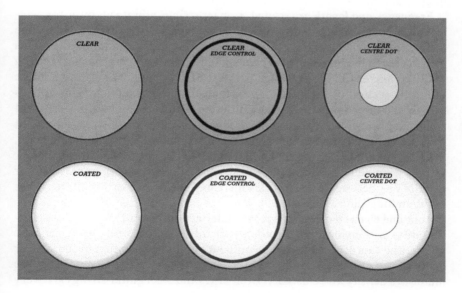

Figure 7.1 Examples of different drumhead designs and features.

kit, but nothing comes close to influencing the sound as much as the drumheads you choose and the way you tune them. There are so many choices, as shown in Figure 7.1. Clear, coated, single-ply, double-ply, built-in overtone vibration control systems, centre dots, and combinations of all the above, not to mention different materials, designs, and manufacturing approaches preferred by different drumhead makers.

7.1   Guitar strings on steroids!

The drummer's drumheads can be thought of as "guitar strings on steroids"! The basic principles for guitar strings apply very similar to drumheads, though with far more options and possibilities for personalisation. In fact, it's a good approach to first understand a few simple concepts about guitar strings and then see how these still relate when considering drumheads.

There are a few basic principles to guitar strings that apply directly to drumheads too, which are as follows:

Frequency and tension are related. The more the guitar string is tensioned or tightened, the higher the frequency of its vibration and hence the higher the pitch of its sound. We see a similar effect with drumheads: the pitch goes up as tension is applied.

Frequency and thickness are related. Thicker guitar strings vibrate at lower frequencies – you can see this on a standard guitar and when comparing the string thickness of a bass guitar to a standard range guitar. This is true also for drumheads; thicker drumheads naturally vibrate at lower frequencies. If two different thickness drumheads are the same size and tightened to the same tension, the one with a thicker drumhead will vibrate at a lower frequency and hence have a lower musical pitch.

Frequency and length are related. Longer strings vibrate at lower frequencies and shorter strings vibrate at higher frequencies. Again, you can see the difference in string length between a bass guitar and a standard guitar. This also relates to the basic musical concept of guitars – playing the guitar string at different frets on the guitar neck is effectively changing the length of the vibrating section of the string, and hence changing the pitch. This applies exactly the same for drums, where different diameter drums allow different frequency ranges for the drum. So, small diameter drums can vibrate at higher frequencies and larger diameter drums vibrate at lower frequencies. With this in mind, a drum set has a number of drums to allow the drummer to play a range of different pitched sounds in their performance. A guitar usually has between four and seven strings, whereas a drum kit often has between four and seven drums.

Tone and character is related to the material used. We all know that a classical guitar with nylon strings sounds very different to a steel-strung acoustic guitar, and this aspect has a huge influence on our perception of the style or genre of the music that we hear. More scientifically, the difference is in the material's density and elasticity. Even brass and steel guitar strings give very subtly different character to the sound, owing to slight differences in density and elasticity. This applies quite similar for drumheads; manufacturers use their own unique formulas to make drumheads (mostly out of polyester), and hence, each manufacturer's drumheads can give a subtly different tone.

A pure vibration profile is related to engineering accuracy. This means that a string which has exactly the same density, thickness, and elasticity along its whole length will vibrate with more pure frequency profiles and with perfect musical harmonics. It might sound obvious that a string will be manufactured consistently, but even if they are, they deteriorate over time. After a few weeks of being played, being strummed and picked in different places with a hard plectrum, by being bent out of shape at certain positions on the fretboard, and being stretched for retuning on a daily basis, the string becomes non-uniform and eventually its harmonics start to become dull or slightly misaligned. This applies exactly the same for drumheads when considering the consistency or evenness of the drumhead's vibration profile, as referred to previously in Chapter 4. So, the equivalent of a guitarist claiming that their strings are dull or past their best is very similar to a drummer who notices that their drumheads are not equalised or uniform and hence do not give off a perfect smooth tone. Of course, drumheads deteriorate over time too, so it's harder for old worn drumheads to vibrate with a smooth even profile also.

The word *tension* crops up a few times in the list above; most people know what that means, but it is well worth defining properly and understanding a few subtleties related to the term. Tension is a measure of force that is applied to stretch something between two or more points. If we apply too much tension to something, it reaches its failure point and breaks. This applies equally to guitar strings and drumheads, as anyone who's over-tuned a string or drumhead until it breaks will know this well! When you hit the drumhead with a stick, you momentarily apply some greater tension force, and as a result, there is a momentary increase in pitch, sometimes referred to as *pitch glide*. This is an interesting concept particularly for heavily damped drums and hard hitters, but if a drum is hit soft and allowed to ring a little, the pitch glide is fairly unnoticeable to the human ear. Another point on tension is that we can't predict what exact tension will result in what exact frequency, because of all of the other variables in the mix; thickness, density, material, and size. So, even if we can measure the tension of the drumhead, it is not a greatly valuable measure in itself for tuning drums, since we are only really interested in what the drum sounds like. That's why it's really valuable to develop your hearing and understanding of drum sound and to understand how this relates to the acoustics of the drum, so that you can tune drums purely based on what they sound like and how they respond acoustically. In the same way, we really don't care what the tension in a guitar string is, since every guitar string has a very different tension depending on the size of the guitar, the string material, and thickness of the string. As with tuning all musical instruments, we are generally most interested in the frequencies and sound of the string, and as long as it is tuned to the correct frequency for playing with the rest of the band, then it doesn't really matter what the actual tension of the string is. In fact, many guitarists choose different thickness strings to allow them to tune to the necessary frequency with subtly different tension. This means, with thinner strings, guitarists can more easily bend the strings during a rock-and-roll guitar solo and add exciting vibrato and dissonance to the performance. Conversely, guitarists who play predominantly chords and strummed notes often prefer thicker strings at higher tension, which give off a different tone that is often described as warm and full. Either way, many different strings (and drumheads) can be tuned to a similar pitch, so the musician themselves can decide through their choice of strings/drumheads what extra sonic qualities they want to bring into the performance and overall sound of the instrument.

With drumheads, there are a number of additional features that have no relationship to guitars, and hence make drums much more customisable and, unfortunately, also much more difficult to tune. Firstly, a drumhead has multiple tuning points around the drum, and as we discussed earlier with reference to lug tuning, these need to be uniformly tuned in order to allow the drumhead to vibrate smoothly and evenly, and hence to give a clear and smooth tone to the drum sound. Secondly, with popular cylindrical drums, we usually have two drumheads, which multiply the complexity tenfold or more! So, as a drummer, it's valuable to explore and experiment with the thousands of different options. After doing this experimentation over the course of a few years, you will have very good hearing skills, and you'll be able to quantify the sounds of different drums and drumheads with a strong knowledge of the acoustic principles that relate to them. Having read up on the theory, you might want to try certain things based on a personal style or a particular sound you are looking for with a specific drum, and maybe try a few different drumheads that you might consider and cross-compare to make your final choice of preference. Drumheads aren't cheap, but it's worth investing in a good set or two and doing this experimentation yourself. There is no learning like *active learning*; unfortunately, you can't just read it in a book or blog; you have to put it into real practice to get the full benefit!

7.2 The drumhead equation

There's a very famous acoustics equation that defines the frequency or fundamental pitch of a guitar string, called Mersenne's law – we won't go into it here, but do look it up if you're interested in this sort of thing. No surprise, there is a very similar law that relates frequency and pitch to the physical properties of drumheads too (or *circular membranes* as we call them in the world of acoustics). The drumhead relationship is described really well, though into quite deep physics theory, by Fletcher and Rossing in their book *The Physics of Musical Instruments*,[1] but we can break it down here and see how it becomes useful for drummers and sound engineers to know about.

The *drumhead equation*, as we might call it, looks like this:

$$f = \frac{k}{d}\sqrt{\frac{T}{\rho \cdot t}}$$

where f is the drumhead's fundamental vibration frequency (i.e. its pitch), T is the tension, ρ is the drumhead density, t is the drumhead thickness, and d is the drumhead diameter. The value k is just a number, 0.7655 to be precise if using standard units for all of the other variables.

So how is the drumhead equation useful to drummers and sound engineers? Well, it explains a number of things about drumhead design – it's the science that explains what we might already know from listening and experimenting with drumheads. Firstly, we see from the equation that if the value of T increases or were to be given a bigger value, then the value of f on the other side of the equation will increase too. So increasing the tension, i.e. tightening the drumhead with a drum key, causes the drumheads vibration frequency to go up. Well that was obvious, all drummers know that! Secondly, we see that all the other variable terms are on the bottom, underside, or denominator of the fractions in the equation, so if any of those values increase, then the result of the equation (frequency) will decrease. So, a bigger diameter means the drumhead frequency is lower – we knew that already, since bigger diameter drums give lower pitch drum sounds.

Specifically relating to drumhead design, we see that if the drumhead is thicker, the frequency of the drumhead is lower. So thicker drumheads give a lower frequency, and hence sound, as some might be aware, a bit "warmer" or "deeper", whereas thinner drumheads have a bit more "brightness" and can be tuned to higher frequencies. In fact, rather than having thicker drumheads, often the thickness is provided by having two thin drumheads sandwiched together, known as *double-ply* or *2-ply* drumheads. These allow the benefits of thicker drumheads without actually having to make the drumhead material so thick that it becomes harder to vibrate properly. So you can simply think of *single-ply* (or *1-ply*) drumheads as being "thin" and double-ply drumheads as being "thick" – because the scientific principles apply very similarly. We also see that the density of the drumhead makes a difference, and a higher density material (i.e. with a higher ρ value) also gives a lower frequency. So it's no surprise that coated drumheads, which essentially are made heavier and more dense by the coating, also sound a bit "warmer" or "deeper", whereas uncoated (less dense) drumheads have a bit more "brightness" and can be tuned to higher frequencies. Thicker and more dense drumheads tend to be more durable and resilient to hard hits too, so there are practical reasons for the different drumhead designs as well. The core science relating to drumheads therefore explains a number of things we intrinsically know from experimenting with different drumheads, listening for ourselves, and of course matches what a number of drumhead manufacturers will say about their different drumhead types when helping you to choose the most appropriate kind for your preference. But there are a number of other factors with drumheads too, including the relationship between the overtones and the fundamental frequency, which is something we always come back to discussing.

We know the fundamental frequency relates to the overall pitch of the drum, and it's excited most when the drum is hit in the middle. And we know that the overtones are excited more when we hit the drum at the edge. But in reality, all of the frequencies are vibrating all together at the same time, and each drum has a slightly different balance of energy between the fundamental and overtones. This explains why some drums give a deep thud sound with not much brightness, whereas others seem to ring on forever with a high-frequency shimmer, lasting long after the main drum sound has decayed away. Different types of drumheads give a very different balance between the fundamental and overtones that we hear, often owing to the thickness or coating, but also drumhead manufacturers have learnt how to manipulate this relationship and create drumhead designs that allow different ranges of overtone control. In particular, many drumhead designs have overtone dampers built in around the edge of the drumhead to stop the overtones from vibrating too long. So if you prefer your drums to be low, boomy, and powerful, you should perhaps go for a coated drumhead with overtone control built in. If, on the other hand, you like a bright, rich sound that is more "musical" and with a longer decay time, then go for a thin clear drumhead without overtone control built in.

7.3 Drumhead types and features

So there are many different options and various innovative approaches by drumhead manufacturers. Let's break these down into categories with a little summary on each.

Single-ply or double-ply

As mentioned above, single-ply or thinner drumheads tend to have a bright character and can be tuned to higher frequencies, whereas double-ply or thicker drumheads give a lower frequency, and hence generally sound "warmer" or "deeper".

Coated or uncoated

As with thin and thick drumheads, clear drumheads tend to have a bright character and can be tuned to higher frequencies, whereas coated drumheads give a lower frequency, and hence sound "warmer" or "deeper". Coated drumheads give another sonic characteristic which is to do with the drumstick contact; because the coating is a harder material than the standard polyester drumhead material, the stick contact is much more like a contact between two hard objects. As a result, the contact time is reduced and this gives the result of what we call a fast *attack*; it's a more crisp sound of stick contact which some people like. A faster attack can be compared to the difference between a hard contact of a wooden beater on a metal glockenspiel and a softer impact by a felt mallet on a wooden xylophone. So clear drumheads have a slightly softer attack sound to coated drumheads, which have a more "crisp" sound when the stick makes contact. This is one reason why snare drums are usually set up with a coated drumhead, as well as the fact that the coating brings a bit more durability and enhances the drumhead's lifespan too. Note that there are many different types of coatings by different manufacturers, some drumhead coatings are thicker and heavier, some are more lightweight, transparent and described as "frosted", and some even use synthetic felt or fibre coatings to give a retro nod to the days when animal hide was used for drumheads.

Edge control

Edge control is essentially the addition of some form of mass, extra thickness, or other design which is intended to reduce the overtones that are excited at the edge of the drum. While we analyse the first overtone frequency to tune the drumhead and ensure the head vibrates evenly, many drummers do not like the sound of the overtone ringing on longer than the main fundamental frequency of the drum. Drumhead manufacturers use different techniques to achieve the same result – some manufacturers use a built-in damper ring, some have an extra ply of material around the edge of the drum, and some even sandwich a little bit of fluid or glue in between each of the drumhead plys at the edge. Whatever the approach, the point is that these drumheads have a reduced overtone sound in comparison with the volume of the fundamental pitch of the drum. It has to be said, though, that overtones are not a bad thing; moreover, they suit some styles of music more than others. Generally, you'll find jazz drummers quite like to hear the drumhead's overtones, whereas rock and metal drummers prefer overtones to decay quite quickly and let the lower fundamental frequency stand out alone. But there are no rules and there are many great drummers who have completely opposite approaches.

Centre dot

The centre dot is almost the same principle as the edge control, but focused on the centre of the drum rather than the edge, and predominantly designed for snare drums. The centre dot is an extra thickness or mass at the centre of the drumhead, which changes the vibration of the fundamental frequency. The extra weight of the centre dot causes the drumhead to stop vibrating after a shorter time, so the sustain of all frequencies is reduced by a similar amount. The dot also gives an increased attack to the sound, similar to an extra thick coating, and hence gives a crisp sonic "focus" and increased durability. Centre dot drumheads tend to be used more often on snare drums, to reduce the sustain of the sound and to enable hard hitting drummers to avoid replacing their snare heads too often!

Of course, you can have combinations of the above too, so the possibilities are almost endless, and make choosing drumheads a fun but virtually limitless task. Table 7.1 shows some different feature combinations for tom drums, and some examples of each type by three of the main drumhead manufacturers: Aquarian, Evans, and Remo.

The drumhead concepts relate similarly for snare drums too. Usually coated drumheads are used on the snare, to improve the attack or focus of stick hits and also because they are more durable and last longer. Many drummers choose to use a centre dot type or one with edge control to dampen the fundamental and/or the overtone frequencies of the drum. Drummer Alex Reeves gives a valuable insight into his drumhead choices for the snare drum:

> My preference right now is for my snares to have a medium thickness coated head with a dampening dot in the middle - in my case I use an Aquarian Texture Coated head with Power Dot. Dampening the middle rather than the edge of the snare drum has a lovely controlling effect on the drum's overtones - quite different to using a dampening ring, tape, or damper gel further to the edge of the drum head.[2]

Table 7.2 shows a number of drumhead examples for snare drums (all coated drumheads).

Table 7.1 Drumhead choices for tom drums, with examples by major manufactures

Ply	Coated	Overtone Control	Example Drumhead
Single	–	–	Aquarian Classic Clear
Single	Yes	–	Aquarian Texture Coated
Single	–	Yes	Remo Powerstroke P3 Clear
Single	Yes	Yes	Remo Powerstroke P3 Coated
Double	–	–	Evans G2 Clear
Double	Yes	–	Remo Emperor Coated
Double	–	Yes	Remo Pinstripe Clear
Double	Yes	Yes	Evans EC2S Frosted

Table 7.2 Drumhead choices for snare drums, with examples by major manufactures (all coated drumheads)

Ply	Overtone Control	Centre Dot	Example Drumhead
Single	–	–	Aquarian Texture Coated
Single	Yes	–	Evans Genera Coated
Single	–	Yes	Remo Controlled Sound Coated Black Dot
Single	Yes	Yes	Aquarian Hi-Energy Snare Head
Double	–	–	Evans Super Tough Coated Snare
Double	Yes	–	Remo Pinstripe Coated
Double	–	Yes	Aquarian Hi-Velocity Power Dot
Double	Yes	Yes	Evans EC Reverse Dot

TRY FOR YOURSELF: DRUMHEAD EXPLORATION

Every now and then it's worth investing in some different drumheads and spending some time evaluating what you do and don't like about different types. Drumhead manufacturers are always developing new designs, and you can only really decide if they are right for you with some thoughtful evaluation in the context of your own drum kit – it's not really possible to decide based on an online video or a quick few hits in a shop. You really need to experiment and compare, which is a great learning process for your hearing skills too.

If you don't want to spend too much money, start with just one drum. Buy three drumheads of different types for a single drum size, a rack tom is a good choice to keep cost down. You can choose which to buy and try based on the information in this book and some on-line recommendations. Tune up one of the heads following the holistic procedure from Chapter 6 (but don't add any extra damping!). Now take a reading of the F0 and F1 values with iDrumTune and write down some things that you do and don't like about the sound, you may even want to take a quick recording of the drum sound with a microphone, if you have a home or semi-professional recording set-up available. Now try the second drumhead and tune it to the same F0 and F1 frequencies as the first. Repeat for the third drumhead too. It's valuable to be able to articulate what you do and don't like about different sounds, and if you have made some quick recordings, then it can be much easier to evaluate between them. (If you are making re-cordings, don't get too close with the microphone for this exercise; it's valuable to record the sound of the whole drum from a distance of 2–3 feet away, similar to the distance of your ears when playing the drum.)

It's very much worth repeating this exploration for your snare drum too. The experience of listening and defining the sounds you hear will help you to become much faster and more capable of achieving a par-ticular snare sound that you might be looking for in future perfor-mance and recording scenarios.

7.4 Resonant drumhead selection

The above theory all relates equally for the resonant drumhead as well, so, with the resonant drumhead, you can experiment just the same. Many drummers tend to use a lighter (thinner) drumhead on the resonant side because it allows an additional range of tonality to be incorporated into the drum sound. We tend to use thicker heads on the top also to avoid excessive high-frequency (overtone) ringing that thinner drumheads give, but it's ok to use a thinner head on the bottom because they don't generally ring out as much as the head that is being hit. It is possible to use identical heads on the top and bottom, but that can limit the frequency range of the drum and also introduces the potential for beat frequencies if the two heads are tuned similar but not exact (similar to the beat frequencies we identified when discussing lug tuning previously). Furthermore, if you find your resonant drumhead is giving off a sustained overtone that you just don't like the sound of, a double-ply, coated, or edge control drumhead on the underside might be what you are looking for.

TRY FOR YOURSELF: EVALUATING RESONANT DRUMHEADS

Using a standard clear resonant drumhead, tune one of your drums to a point where you are happy with the sound – following the holistic process from Chapter 6. Make a note of the F0 and F1 readings with iDrumTune. As before, you may want to make a quick recording of the drum sound for future reference too.

Now change the resonant drumhead to one which is either coated or with a bult-in overtone control feature. Tune the drum to the same F0 and F1 frequencies as before and listen to the drum. How much different does it sound with the heavier drumhead than with a clear resonant drumhead? Can you articulate the difference in sonic terms? Is it a significant difference?

It's worth trying this experiment just to have an understanding of how big a difference the resonant drumhead can make. It's much more subtle than changing the batter drumhead, so you can decide if it's worth the extra cost of more expensive drumheads on the resonant side.

Once you have a good understanding of how the resonant drumhead affects the overall drum sound, you might also want to experiment with more advanced tuning approaches for the resonant head. A thin resonant drumhead tuned with a higher frequency overtone works generally very well and is relatively easy to achieve. However, for example, many drummers like to use a resonant drumhead that is tuned fairly loose in comparison with the batter head, which might be tuned quite high. This allows a sharp attack to the drum sound from the tight batter head, whilst maintaining a warm low fundamental from the relationship between the batter and resonant heads, resulting in musical and impactful drums that still have a strong low-frequency power.

7.5 Experience drumheads!

With drumheads, there is so much information out there and lots of different explanations, terms, and theories, and not enough time to try them all out! We suggest you consider the options above to decide what kind of sound you are aiming for, and then experiment from a best first estimate of what drumheads you require to achieve your sonic objective. There are some great videos online which show comparisons of different drumhead types, particularly those by Drumeo[3] and Sounds Like a Drum,[4] but there really is no replacement for experiencing different drumheads in the context of your own drum set-up and with your own ears. Emre Ramazanoglu (drummer, producer, and recording/mix engineer for Noel Gallagher, Kylie Minogue, and Richard Ashcroft) emphasises this approach and suggests some drumhead choices to consider as a starting point of reference:

> The drumheads make a much bigger difference to the sound than the drum itself. There's not a very good shortcut, you just have to try lots of different kinds and experience what sound you can achieve with each type. But, if I was to supply backline to a festival, without any guidance other than it being rock or jazz, then I'd go for single-ply coated heads for the jazz kit and Remo Powerstroke 3 or Pinstripe heads for the rock kit, which would give a good starting point for most drummers.[5]

So don't be afraid to invest in good drumheads and try a few until you find your perfect type. The drumheads really do have the most influence on your drum sound than even the drums themselves; it's not uncommon for a cheap old drum kit with well-chosen and well-tuned drumheads to sound really great, whereas it is very possible to make an incredible expensive drum kit sound horribly wrong with low-quality drumheads that are poorly tuned.

Notes

1 *The Physics of Musical Instruments* by Neville H. Fletcher and Thomas Rossing (2005), 2nd Edition, Springer, p. 75. Note that the variable σ in Fletcher and Rossing's equation in Figure 3.6 is the "areal mass density" or "mass thickness" of the drumhead (with units of kg/m^2), which is equivalent to the product of density and thickness.
2 Interview with drummer Alex Reeves conducted on 07/10/2020.
3 For example, *How to Choose a Snare Drum Head*, YouTube video by Drumeo, available online at https://youtu.be/kKTBsC_L8kE [accessed 01/08/2020].
4 For example, *Coated Resonant Heads for Natural Tonal Control*, YouTube video by Sounds Like a Drum, available online at https://youtu.be/8PdTEkF0mpc [accessed 01/08/2020].
5 Interview with drummer and music producer Emre Ramazanoglu conducted on 14/10/2020.

8 Timbre
The truth about drum shell vibration

With all musical instruments, there's a hidden quality that makes some sound "better" or different to others, or some instruments just sound more suited to a certain style, genre, or even musician. Often there's no right or wrong; some pianists play a Steinberg, and others prefer a Yamaha. Some guitarists play a Gibson Les Paul and others choose a Fender Stratocaster. But these instruments can all be tuned the same, to give the same frequencies when played and hence to allow a consistent and optimal sound every time. So if these instruments have identical tuning, what is it that makes them sound different? Well, that is all the other magical qualities of the musical instrument, which make some instruments sustain notes with a long clear tone while others sound dull. Some instruments have rich overtones and harmonics, yet others produce sounds more like pure sine waves. Drums, just like any other instrument, are a great example of this.

Drummer and producer Emre Ramazanoglu has valuable experience when comparing different drums kits with similar setups:

> When I was making drum samples for FXpansion, I had the opportunity to use 20 different drum kits all in the same room and all with the same heads. Some kits have a special unique quality to them. The Pearl Reference had a strong attack to the sound and the 70's Jasper shells were just beautifully musical and focused. But it's surprising that many drum kits sound virtually identical with the same heads on too. I think it has a lot to do with how resonant the drum shells are and that can influence how easy they are to play too.[1]

8.1 Introducing timbre

We know that all musical instruments sound different, and drums are a great example of this fact, given the multitude of options for drum designs, drum shell materials, diameters and depths, the metal hardware fixed on the drum, and of course the type and design of the drumheads themselves. All these aspects influence and contribute to the resultant drum sound, in the same way that an acoustic guitar with nylon strings sounds very different to a solid electric guitar with steel strings. In fact, there are three distinct categories of sound which combine to define the overall sound of a musical instrument:

- The tuning of the instrument
- The way the instrument is played
- Everything else!

Actually, the above list is a little provocative, because surely we can't define *everything else* as a scientific or artistic category of sound, can we? Well, actually, that is what acoustics engineers and instrument manufacturers have always done, though more specifically we use the term *timbre* to define everything else that cannot be described by the fundamental and related frequencies of a sound (i.e. the sound's pitch or tuning) and the loudness profile of the sound, which is controlled by the performer. David Howard and Jamie Angus, in their seminal textbook *Acoustics and Psychoacoustics*, describe timbre by the following statement:

> two sounds which are perceived as being different but which have the same perceived loudness and pitch differ by virtue of their timbre.[2]

With this in mind, a slightly more suitable definition of *instrument sound* is given by the following simple equation:

$$instrument\ sound = tuning + performance + timbre$$

Looking at the above equation a little more reflectively, we know that most instruments can be tuned in a huge number of ways. There are many different tuning choices when playing a guitar, from standard *open tuning* to *drop-D tuning* and many other combinations, not to mention the fact that the tuning changes completely on all strings when a capot is used. Even pianos are tuned subtly differently, sometimes on purpose to the taste of the musician or the piano tuner, or unintentionally as it drifts out of perfect tuning over time.

The sound of the performance depends, of course, on the instrument type and the arrangement of notes that are performed, and how they are performed with respect to a time signature or temporal pattern. But in terms of the acoustic qualities of the performance, this predominantly relates to the dynamic profile of the performed sounds, which can be soft, loud, and quiet, with a sharp attack or a sustained release time. Instruments, such as trombone and guitar, allow the performer to add fast attack profiles, held sustain, as well as vibrato, pitch bend, and dynamic slides between musical notes. As you know, there are also many different ways to create sound from a drum with a drumstick and some controlled arm and hand movement, particularly by varying the volume and dynamics of different notes and phrases.

So "everything else", or the timbre, of the sound describes all the other sonic characteristics that we can't define in the aforementioned tuning or performance categories. This is often subjective and qualitative – using descriptive semantic terms, such as rich, warm, mellow, dull, bright, harsh, sombre, and colourful – since we can't easily associate these elements of the sound with clear scientific theory or quantified measures. Some of these qualities can be controlled subtly by the performance, but in general the instrument has its own unique timbre characteristic (or range of timbre characteristics) which are inherent in the instrument design and construction itself. Timbre therefore is a strange term, because it is defined best by discussing what it *isn't* rather than what it *is*, and it is also very open to interpretation, since one person might utilise descriptive terms in a different way to someone else. But with all that in mind, it's a very useful term to help us discuss how two seemingly similar instruments differ in their sonic characteristics, which we experience all the time in the world of music. For example:

- If a musician plays the same chord in the same way on two different guitars, the timbre is what defines the difference between the two sounds.
- A soft middle C note on both a trumpet and a clarinet differs only by the timbre of the sounds.
- The difference between the sounds of two vocalists both singing a forte A4 note is predominantly from the difference in the timbre of their voices.
- Two drums tuned to exhibit the same fundamental and overtone frequencies differ in sound by their timbre.

It's important to recognise that there is a little overlap between "timbre" and "performance", given that it is possible to play an instrument in many different ways, and these performance nuances can enhance or attenuate different timbral qualities of an instrument. So, in many respects, we have two categories of timbre: the more general timbre of *the instrument* and the timbre of *the sound of the instrument* in a particular performance instance. Focusing particularly on the instrument itself, we know that the timbre of a drum describes the sonic differences between two drums that are tuned to the same frequencies and are played in the same way. So we can dig a bit deeper to understand what physical things contribute to the timbre of a drum. Going back to our original list at the start of the chapter, the timbre of drum sound is contributed to by:

- Size and shape of the drum
- The material and design of the drum shell
- The construction and design of the hardware attached to the drum (including lugs, tuning rods, hoops, and any snare wires or other attachments)

- The way the drum is mounted or the stand it is positioned on
- The drumheads that vibrate and generate sound from the drum when hit

The drumheads are, of course, very important and influential, and we've discussed the timbre associated with drumheads separately in Chapter 7. So predominantly here, we'll take a look at the acoustics theory and creative practice relating to the drum shell, the hardware fixed on the drum shell, and the way the drum is mounted.

8.2 Tuning fork example with mass loading

Before we start to consider the frequencies of a drum shell, it's valuable to cover one key acoustics theory before proceeding, and that is about how adding weight and mass affects the frequency of an object or vibrating system (i.e. a system being a number of objects connected together). It's possible to evaluate this effect with a tuning fork that has some additional mass added to it. Figure 8.1a shows the frequency spectrum of an F4 tuning fork when measured by iDrumTune. As expected, it has a single peak at the F4 frequency of 174.0 Hz. When a small, arbitrary amount of mass is added to one of the tuning fork tines, the frequency reduces to 160.5 (Figure 8.1b), and when some mass is added to the other tuning fork tine, its vibration frequency changes again to 146.0 Hz (Figure 8.1c).

With additional weight added, the tuning fork no longer vibrates at its pure musical frequency F4, and it would be totally useless for using as a reference pitch if it were used with some weight added. Figure 8.1 also shows that the frequency peak becomes less defined as a single clear spike as we add more mass, since the added mass is also restricting the tuning fork from vibrating as it is designed to, with a pure sinusoidal profile. The same applies with removing weight: if we were to file the tuning fork down or hacksaw a bit off its tines, then its vibration frequency would go up. This tells us

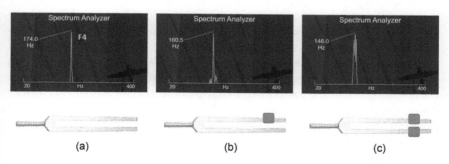

Figure 8.1 Tuning fork frequency spectra for (a) free vibration and (b and c) with different amounts of mass added to the fork tines.[3]

that we need to consider the system as a whole single vibrating system, and it's not possible to just consider the tuning fork's own frequency, because as soon as we add anything to it, that frequency is gone and irrelevant, it doesn't exist anymore. With weight added, the tuning fork has a new fundamental frequency altogether.

TRY FOR YOURSELF: TUNING FORK LOADING

If you have a tuning fork or a glockenspiel bar, you can measure its vibration frequency with iDrumTune. It's best to use a soft mallet and hit the bar or fork repeatedly but quite softly. Aim to get the bar vibrating quite consistently and continuously with strokes or taps with the mallet. Now hold the iDrumTune microphone close and see if you can just get enough volume to log a reading in the Spectrum Analyzer mode.

Now attach a small piece of metal to the end that is vibrating freely – use some tape to hold it in place. Cause the bar or tuning fork to vibrate again, and you'll see that the main fundamental frequency has reduced by a small amount. Change the position of the weight or add some more weight, and you'll see the vibration frequency change again.

8.3 Drum shell vibration

A complete assembled drum is very similar to the tuning fork with weight added. When assembling a drum, we start with a perfect clean and round drum shell, and then add extra mass in the form of lugs, hardware, hoops, tuning rods, and drumheads to the drum shell. We can hence conduct the same experiment as with the tuning fork, but relating to additional mass added to a drum shell.

We saw back in Chapter 2 that, when hit, a drumhead vibrates at a very specific fundamental frequency with related overtones, and each frequency and overtone has a related vibration shape (or *mode shape*). As with all physical things, a drum shell will also vibrate at very particular frequencies and mode shapes when excited. However, the drum shell vibrates in a different direction to the drumhead; in fact it vibrates perfectly perpendicular in a horizontal (as opposed to vertical) axis. So as the drumheads vibrate up and down, the drum shell vibrates from side to side. Figure 8.2 shows the most fundamental vibration mode of a circular drum shell. As the shell vibrates, its shape deforms from a perfect circle into a slight oval shape with a greater radius in one direction. Then as it vibrates back towards its start position, it overshoots and deforms with a slightly shorter radius and continues back and forth until eventually all vibration energy is lost and the shell returns to its normal rest position and shape. The shell therefore vibrates in and

Figure 8.2 Drum shell deform-reform vibration profile (fundamental mode).

out like a circle that gradually deforms and reforms periodically. Figure 8.2 shows how most circular structures vibrate when struck on axis, though with more detailed analysis, we do also see further deform-reform shapes relating to less powerful overtone modes.[4]

When the drum is struck on the drumhead, the drumheads vibrate up and down, but all that energy moving around causes all the other components of the drum to vibrate too, so the drum shell and the metal hoops and hardware all vibrate too. These vibrations are far lower in energy than the drumheads themselves, and they do not really change based on the material or tuned frequency of the drumheads. So the drum shell vibration doesn't contribute to the pitch or tuning of the drum; it just contributes to the more overriding timbre of the sound, i.e. the extra little qualities that make two identically tuned drums sound different. We can see this often on the frequency spectrum chart of a drum hit. For example, Figure 8.3 shows the fundamental frequency of the drumhead vibrating up and down, as well as the overtone vibration mode of the drumhead. Sometimes we also see some extra little bits of frequency data at the higher areas of the spectrum, which are more likely related to the drum shell and the metal hardware. In fact, the drum shell and the metal hardware are generally vibrating at much higher frequencies than those related to the tuning of the drumheads. This

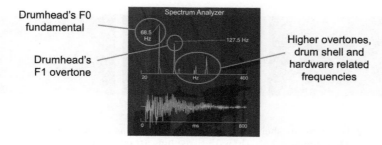

Figure 8.3 Identification of a drumhead's fundamental and overtone frequencies, alongside additional timbre-related energy from the drumhead's higher overtones, the drum shell, and the attached hardware.

is especially apparent when the whole drum is constructed together with hoops and mountings doing their best to stop the drum shell vibrating at all. Additionally, heavy solid materials that are more dense and less elastic than drumheads have higher natural frequencies by nature, so the timbre aspects of drum sound, provided by the drum shell and attachments, are generally significantly higher up the frequency spectrum than the frequencies we are interested in with respect to tuning the drumheads.

8.4 Loading the drum shell

It's valuable to understand the acoustic difference between a blank, perfectly round and untouched finished drum shell, and a drum shell as it appears within a working drum kit, with metal hardware, hoops, drumheads, and other mounting and hardware attachments. While it's interesting to evaluate the vibration frequencies of a simple and pure drum shell, this analysis has no major relevance to the acoustics of a complete drum shell as part of a practical drum in a performance scenario. Every change that is made to the drum shell, and every bolt and component that is added, loads the drum shell and stops it vibrating at its pure natural frequency, and as a result the drum shell vibrates at a totally different frequency instead.

In an experiment with a six-lug, 13″, five-ply birch drum, it was seen that the vibration frequency of the shell changes considerably with different amounts of hardware attached. It's possible to hold a bare shell with just your fingertips and gently cause it to vibrate with a mallet, and then use a microphone and spectrum analyser (such as that in iDrumTune) to measure the vibration frequency. The bare shell of the drum used in this experiment was measured to have a fundamental shell vibration frequency of 89 Hz, with an overtone vibration frequency at 252.5 Hz (as shown in Figure 8.4a). When adding the tuning lugs (i.e. extra weight), the shell starts to vibrate at a lower frequency when hit by the beater. The shell vibration frequency becomes 79.5 Hz with three lugs attached (Figure 8.4b) and 73.0 Hz with all six lugs attached (Figure 8.4c).[5]

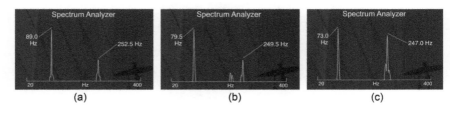

Figure 8.4 Birch drum shell vibration frequency measurements for (a) the bare 13″-diameter shell, (b) with three lugs attached, and (c) with all six lugs attached.

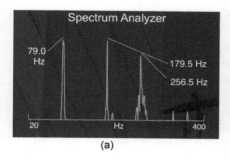
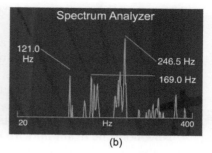

(a)   (b)

Figure 8.5 The 13″ birch drum shell with (a) all lugs and one hoop and (b) with both hoops added.

Adding hoops makes a huge difference too, because they not only add more mass, but they are also very rigid and actually attempt to stop the drum vibrating in the deform-reform profile. This results in a lot of new frequencies occurring in the drum as it approaches that of a complete drum which we would play in the drum kit. With one hoop added, the drum in this experiment vibrates at a fundamental frequency of 79 Hz, and some additional vibration frequencies appear owing to the vibration of the hoop itself, as shown in Figure 8.5a. It gets harder to measure the shell's vibration frequency as we add more and more things, but we know one thing for sure, the drum's vibration profile and hence the drum's sound are very different with all the hardware attached. With the addition of the second hoop, there is no trace of the shell vibration frequency at all, as shown in Figure 8.5b. The hoops have stopped the drum shell from vibrating with an amplitude (volume) that is big enough to measure. We see lots of new frequency peaks, because the drum is now a complex vibrating system, and you can certainly hear the metallic vibration of the hoops themselves too.

With the fully assembled drum, the shell is, of course, still vibrating and it is still providing a component to the sound of the drum, but it is very small and subtle in comparison with the volume of sound that is generated by the drumheads vibrating. This is why we can tune two drums identically, but they will still sound subtly different; the timbre of the shell, hoops, the lugs, the mounting, and the style and type of drumheads themselves all give the drum a unique sound that is independent of the drum's actual tuning. But importantly, regardless of their shell material or manufacturing method, all cylindrical drums can be tuned low and all cylindrical drums can be tuned high – so the drum shell has no influence on this element of the sound, which is purely to do with the drumheads and how tight they are tensioned, and how evenly they are tuned.

TRY FOR YOURSELF: DRUM SHELL VIBRATION ANALYSIS

If you have an old drum lying around, it's very interesting to take it apart and see how its shell vibration changes from the bare shell to one which is fully loaded with lugs, hoops, and drumheads. Remove all the hardware and the lugs from the drum shell too. If you use a 13" or 14" drum, you should be able to get inside with a small electric screwdriver and undo the attachments that hold the lugs in place.

Now hold the drum up with one or two fingers (so that you don't stop the shell vibrating) and tap the shell gently with a mallet, and you should be able to hear the shell frequency. If you have iDrumTune set-up nearby, it should be able to capture a reading of this frequency in the Spectrum Analyzer. Add half of the lugs back to the drum and take another reading, and you should see the frequency of the drum shell has changed and become lower. Now add the rest of the lugs and take another reading – the frequency should have come down again.

Add the hoops onto the drum (with no drumheads for now), and you'll find they almost stop the drum shell from vibrating at all. The nice tone of the original bare drum shell has gone and been replaced by a whole new sound characteristic.

The drum hoops therefore cause the drum to stop vibrating. In acoustics and vibration terminology, we call these *boundary conditions*. Boundary conditions are conditions which we know have a fixed (or almost fixed) position on the vibrating system. Usually a boundary condition defines the edges of the vibrating system where it is not possible for the system to vibrate at all. With respect to drumheads, the hoops are a boundary condition, because they stop the drumhead from vibrating right at the very edge of the drum, the hoops ensure that all vibration takes place within the drumhead itself, and the drumhead remains fixed or stationary at the exact point of the perimeter of the drum. The hoops cause a boundary condition to the drum shell too, as a point where it's very difficult for the shell to vibrate, because it is held firmly in place by the rigid metal hoops. There is actually a little vibration, but as we saw above, the hoops generally cause the shell vibration to be very small. This small amount is still audible as an aspect of the timbre of the drum, but it is not sufficient to influence the tuning of the drumhead at all.

Another type of boundary condition relates to how each drum is positioned on or within the drum kit. Tom drums sound subtly different when placed on a snare stand in comparison with when placed on a rack mounting held in place over the kick drum. Equally, a floor tom sounds subtly different with its rubber feet removed. Again, these conditions affect the timbre of the drum, and hence have a valuable role in the overall sound of the drum. But, equally, these types of conditions are not significant enough to affect what frequencies the drumhead vibrates at when it is hit.

TRY FOR YOURSELF: EXPERIENCING BOUNDARY CONDITIONS

A simple exercise to show the effect of boundary conditions is with two drumsticks. Hold both drumsticks at the end with just two or three fingers and tap them together. Now move your grip on one drumstick to the center and continue to tap as before. You'll notice the sound has changed based on the boundary conditions you applied to the sticks, i.e. where the sticks were held in place.

You can experiment more with boundary conditions to hear the subtle differences that can be made to a drum kit. Listen to your rack tom attached both above the kick drum and alone on a snare stand. Do you prefer the sound of one or the other? Also try with your floor tom, either remove the rubber feet or turn the legs over so that the metal ends touch the ground. Maybe even try with two legs the correct way up and one upside down. It's very interesting to hear the subtle differences. Of course, no set-up is either right or wrong, but it's valuable to get an idea of what subtle timbre sounds you like and what aspects you can control around your kit.

8.5 Bearing edges and precision manufacturing

No discussion of drum shells would be complete without mentioning bearing edges. The bearing edge is the edge of the drum shell which makes contact with the drumhead and therefore constitutes the boundary of drumhead vibration as mentioned above. The role of the bearing edge is to assist and allow the drumhead to vibrate evenly and smoothly. The bearing edge needs to therefore be perfectly round in order for the same boundary condition to apply at all positions around the perimeter of the drum. If the drum shell itself is not round, then it is unlikely that the bearing edge will be round. However, a perfectly round drum shell may still have an imperfect bearing edge if not machined properly in the factory. It's actually a very difficult task to make a perfectly round drum with a perfectly round bearing edge (and just as hard to measure how round a bearing edge actually is), so this can explain why some drums sound better and tune easier than others. If the bearing edge of a drum is even and round, then any good quality drumhead can be tuned to give an even response profile.

There are a number of common designs for the shape and profile of a bearing edge – some using very sharp points and others with more rounded profiles.[6] The bearing edge is another component that affects the timbre of the drum, since more contact area of the bearing edge with the drumhead causes more energy transition from the drumhead to the drum shell itself. As a result, it's often regarded that sharper bearing edge peaks allow the drumhead to vibrate with more resonance and sustain than those with rounder edges, so this might be an influencing factor of the sound you are aiming for when buying a new drum kit.

It's also important to keep your drums round! The rigid metal hoops are very good for keeping drums round, but, if removed for any length of time, it's not uncommon for drums to become slightly deformed or *out of round*. Thankfully strong hoops can pull a drum back into round, but it's still better for the drum to stay round on its own accord. If you are changing both sets of drumheads at any time, it's an interesting check to remove both hoops and measure the diameter across the drum shell in a few different directions with a tape measure. Hopefully you get the same measurement every time, proving that the drum is indeed round.

Similar applies for hoops, which, of course, should also be perfectly round. Furthermore, if you place a hoop on a flat surface, it should sit flat on the surface for the whole perimeter of the hoop. If not, don't worry too much, because usually the tuning rods are more than capable of pulling the hoop flat on the drum's bearing edges (and hence are still able to give an even drumhead tension), unless the hoop is significantly deformed.

8.6 Considering the drum shell vibration frequency when tuning

Science and acoustics tell us that the vibration of the drum shell is only really relevant when considering the whole assembled drum. It's really not of any major value to consider the vibration frequency of a bare drum shell in terms of performance or tuning, because that frequency changes as soon as you mount or attach anything to the drum – and it's unplayable without lugs, hoops, and drumheads attached! But, of course, the vibration of the drum shell is a valuable contributor to the complete sound of the drum. Thankfully, these things – the drum shell design, the drum material, the types of hardware, and the mounting of the drum, even the shape of the bearing edges – are all related to timbre of the sound and not what we call the tuning of the sound. It's an important distinction to make and is one that explains why two instruments (including drums) can be tuned the same but yet sound quite different.

So the truth is, there is no scientific or creative benefit to considering the vibration frequency of the bare drum shell when tuning the drum and manipulating drumhead vibration frequencies. Any instrument can be tuned and performed in many ways – but it's not easy to change the timbre of the instrument, which is inherent in the design and manufacture of the instrument. You can't change the timbre of a Steinway piano very easily! But with a guitar you can choose to use your preferred type of strings and with drums you can choose different hardware, mountings, and drumheads too, so we do have some control and choices to make with regard to timbre. It's therefore best to keep your thinking about tuning and timbre separate. As with performance, they do, of course, all interact creatively and if you can maximise the quality of your tuning, the timbre of your kit and the quality of your performance, then you're maximising the individual components of the instrument sound equation we gave at the start of the chapter, and you can be sure to be sounding great whenever you sit down to play. We can all tell if a musical instrument is being played well, or if it is tuned badly, but as with much art and creativity, often one person's chalk is another person's cheese with respect to timbre, and that's one of the many wonders of music!

TRY FOR YOURSELF: DRUM SHELL MATERIALS

It's really valuable to try different drum shell materials and designs if you can. Drums can be constructed of thin, light, heavy, and dense woods, all which have subtly different vibration characteristics. Drums can be made of different metals, transparent acrylic and even empty plastic tubs from a rubbish tip. It's hard to know which type of kit to purchase, so maybe try to swap one drum with a friend who has a different kit, just for you to both experiment with. If you have two 13″ drums made from different materials, you can tune them both to the same fundamental and overtone frequencies and listen to the sound – even better if you can try the same type of drumhead on each to give a fair comparison.

You'll probably be surprised how similar many drums sound when they are tuned the same, regardless of their construction material. Drum sound comes very much predominantly from the drumheads vibrating, so the drum shell itself has a much more subtle influence on the sound. But that little sonic difference may be what you love about your drums, so it's worth finding the opportunity to experiment and seeking out your perfect kit type.

Notes

1 Interview with drummer and music producer Emre Ramazanoglu conducted on 14/10/2020.
2 David M. Howard and Jamie Angus, 2017, *Acoustics and Psychoacoustics*, 5th Edition, Focal Press, p. 238.
3 As shown in an example YouTube video by iDrumTune, titled *Music Instrument Vibration – Tuning Fork Example*, available online at https://youtu.be/pizdR3n-l0dc [accessed 01/08/2020].
4 Drum shell vibration modes are discussed in more scientific detail, including holographic interferograms that verify the vibration mode shapes of a drum shell, by Thomas D. Rossing in his book *Science of Percussion Instruments*, 2000, Volume 3, World Scientific Publishing Co., pp. 29–30.
5 As shown in iDrumTune's YouTube videos *Drum Shell Vibration* discussed by Professor Rob Toulson (Parts 1 and 2) available at https://youtu.be/V_CMNGSuXVo and https://youtu.be/0pyTg3B3_NI [accessed 01/08/2020].
6 Bearing edge profiles described and discussed in *What You Need to Know About… Bearing Edge*s in Drum Business Magazine, September/October 2013, available online at https://www.moderndrummer.com/2014/12/need-know-bearing-edges/ [accessed 01/08/2020].

9 Tuning for different musical styles and genres

Taking a seat behind a fully loaded drum kit brings a moment of joy, power, and excitement to all drummers, and even non-drummers too! Just watch someone's face light up as they take hold of the drumsticks and experience the thrill of sitting behind a kit for the very first time. Every drummer's kit has been meticulously set up to enable the perfect performance for that one person – it's an incredibly personal thing. Some drummers have more cymbals than drums, some have two floor toms, some have two kick pedals, some have toms mounted on the kick drum, some have their toms mounted on a cymbal stand or snare stand – the choices are endless. And even if two drummers have exactly the same kit and equipment, their set-up will differ based on the way they play, how tall they are, how high they like the seat, or whether they want the dark thin crash on the left- or right-hand side. What's more, with a single drum kit, physically set up to suit the owner's needs, there are still thousands of choices on how the kit is played and how it is tuned for the type of performance. A drummer might play jazz-style drums in the morning and rock drums at night, or maybe play in two bands and need two different drum kits to suit the style and genre of each one.

You may have heard that drums are often regarded as unpitched or non-musical by many people, which is an argument that doesn't really hold much credibility any more, given the most recent research and understanding. It's often mentioned because people find the sound of drums hard to understand, because they don't have perfect harmonics like, for example, a piano, guitar, or trumpet. But that doesn't mean drums are un-musical or without pitch – cylindrical drums are certainly musical instruments that exhibit discernible pitch. We've also seen that the theories associated with drumhead design and vibration profiles very much lend themselves to fundamental pitch and tuning, despite the fact that drums have predominantly in-harmonic overtones, and it's both essential and empowering to consider the musical pitch of drums when tuning and setting up a drum kit.

Legendary drummer Terry Bozzio (Frank Zappa, Herbie Hancock, Korn) has pioneered approaches towards musical tuning of drums, with his "Big Kit" including 26 toms each tuned to a different musical pitch. Bozzio's kit and tuning set-up is outlined in Figure 9.1, incorporating 14 8″ × 3″

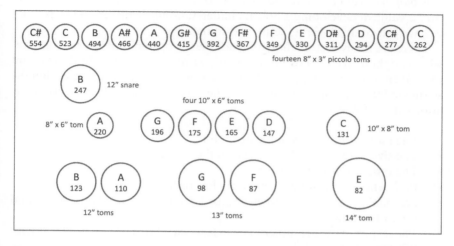

Figure 9.1 Simplified overview of Terry Bozzio's Big Kit tuning set-up (fundamental frequency values given in hertz).

piccolo toms tuned chromatically from C#5 (with a fundamental frequency of 554.4 Hz) down to C4, a B3-tuned 12″ snare, and a range of other 8″- to 14″-diameter drums covering lower pitches from A3 down to E2 at 82.4 Hz.[1]

We know from our previous discussions, and from Bozzio's kit set-up, that it's possible to tune the fundamental pitch of a single drum to a number of different frequencies. But, how do you decide where to tune a drum within this range? Perhaps, more importantly, how do you tune all of the drums in your drum kit so they work well together as a single musical instrument, giving exciting drum rolls around the kit and complimenting the style of music being played?

9.1 Creative objectives

While it's essential for drummers to develop and perfect a personal drumming style, and be flexible to play different styles too, it's just as important to be able to set up and tune a kit relevant to these specialisms or styles too. There are hundreds of different ways to tune and optimise the sound of a drum kit, but some approaches suit certain styles better than others. For example, jazz drummers generally tend to tune a kit to higher frequencies than rock drummers. As mentioned a number of times already, the endless possibilities for personalising a drum kit and its sound are both a blessing and a curse, because your knowledge and experience needs to evolve every time you want to try something new. Drummers and studio engineers should really never stop learning or investigating new options and approaches. It's great to have more than one trick up your sleeve – being able to quickly change the sound of your kit in a studio session or when auditioning for a new band is a valuable skill. So there is no "one-size-fits-all" solution, moreover a value in developing a holistic understanding of the drum kit in order to have full control over its setup and sound.

There are many factors that influence the set-up and tuning of a drum kit, which might be referred to as *creative objectives*. This means thinking about the end creative or musical result and then working backwards to design the set-up and tuning approach to deliver the creative objective. Some considerations are as follows:

- The number of drums in the drum kit
- The size of the individual drums in the drum kit
- The types of drumheads used
- Preferences for tuning to relatively low or high pitches
- Personal preferences between damped drum sounds and more open and tuneful sounds
- The response the drummer likes to feel when hitting the drumhead with the drumsticks
- The type of drum fills and grooves that the drummer plays
- The genre of music that is being played
- The song that is being played and the musical key of that song

It might sound a bit simplistic, but a "rock" drummer would generally play a "rock" kit with "rock" drumheads and with a "rock" tuning. No surprise, but also, a "jazz" drummer will usually play "jazz" drums with "jazz" drumheads and "jazz" tunings. But what does this mean? These are just words that don't really help anyone practically without some details added, and more interestingly, what happens if we use a jazz kit with rock drumheads and set up with a tuning somewhere in between? That could be interesting... or maybe it would be a disaster!

What this type of thinking confirms for us is that what's right for one person, or project, can be horribly wrong for another, and that's ok, in fact, that's one of the really beautiful things about drums. The best approach is to continuously develop your knowledge and understanding, and make a very personal, informed choice about what sound you want from your kit, or the kit of a drummer you are working with in the studio. If you have a good understanding about drum sizes, drumheads, and drum tuning for different music styles and genres, and how they relate to each other, then you are well equipped for many different scenarios that might present themselves.

9.2 Drum sizes for different music genres

The drum sizes chosen for a particular style relate back to the drumhead equation we saw in Chapter 7. This theory told us that if two different sized drumheads are made from exactly the same material and are tuned to exactly the same tension, then one with a larger diameter will vibrate at a lower frequency. So if you want to have drums that are to give deep low frequencies and are tuned to be powerful and bass heavy, then it makes sense to have larger diameter drums. The opposite applies too, so if you want to play drums that sound bright and tuned high, then it makes sense to have smaller diameter drums in your kit. You can have some fun with this theory though, experimenting with your own style and preference – there's no reason why you can't have large drums tuned high or small drums tuned low for example. A great example of a drummer who developed a unique sonic approach is John Bonham of Led Zeppelin, who was known to be significantly influenced by the jazz world, but also wanted to contribute powerful and identifiable drums to rock songs. Bonham's approach to drum sound was therefore to use large diameter drums (often a 26″ kick, 18″ and 16″ floor toms, and a big 14″ rack tom), yet with drumheads tuned to a fairly high tension, giving a relatively high fundamental pitch and a clearly defined sound to each drum.[2] Drummer and music producer Emre Ramazanoglu explains more about how the two most standard approaches for jazz and rock tuning can be combined and personalised:

> It's true that jazz drummers tend to tune high and rock drummers tend to tune lower. But in reality it's more intricate and there are a lot of options. It's still possible to have a deep rocky sounding kit with quite high tunings, if the drums are fairly deep and with low tuned resonant heads. That way you get the attack and up-front tone from the tight batter head and a deep warm sound from the depth of the drum and the lower tuned resonant head.[3]

Over the years, different drum manufacturers have seen what sort of kit sizes drummers tend to like for different styles of music. Of course, there are drummers who like to have just two toms and drummers who want up to eight or more toms, so the number of drums in your kit does influence the choice of sizes to go for. For example, if using two rack toms, you might choose a 12″ and a 14″, whereas if you use just one rack tom, you may choose a 13″ diameter drum. The choice of drum sizes also relates very directly to the range of tuning that you want over the whole kit: some drummers like all their drums to occupy a low range on the kit, others like to occupy a high pitch range, whereas others like to cover a very wide range with some drums giving very low pitch sounds, and other drums giving very high pitch sounds. So the pitch intervals between the drums becomes a consideration too. Every drum should be in the drum kit for a reason – and that reason should usually be to give a sound that none of the other drums give. Most common is for drummers to use a four- or five-piece drum kit, which includes counting the snare and kick drum. So, a four-piece kit usually has one floor tom and one rack tom, whereas a five-piece kit usually has one floor tom and two rack toms (or vice versa). However, it's not uncommon for kits to have two floor toms and three rack toms, and some drummers like to have

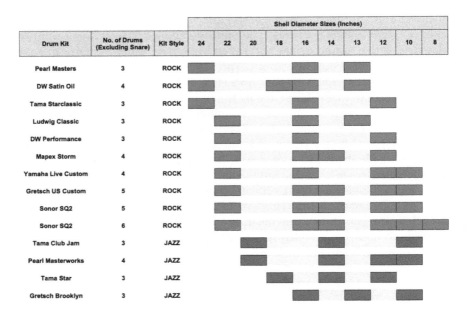

Figure 9.2 Examples of commercially available rock and jazz drum kits (as described by manufacturers), detailing drum sizes in each kit.

a full rack array of different sized toms which can be played like a complex musical instrument.

Figure 9.2 lists a number of different drum kits that are on the market, detailing the range of sizes which kits are sold in. In many cases, the manufacturers cater for many different shell sizes in each kit, so it's possible to create bespoke and hybrid configurations too. A few observations from Figure 9.1 show that, in general, drums marketed as rock style occupy larger diameters, whereas drums marketed as jazz style are predominantly smaller and with less total number of drums. However, for example, the rock-style Sonor SQ2 has matching tom sizes to the Tama Club Jam jazz kit, so it is certainly possible to use a single kit for different styles and genres of music, if set up and tuned suitably in different scenarios.

A quick note about the depth of a cylindrical drum; we've seen that the tuning frequency is related to the diameter of the drum, but what is the influence of the depth of the drum? Well, this really has no major influence on the pitch of the drum; you can tune a deep 16″ floor tom to exactly the same frequencies as a more shallow 16″ floor tom. Of course, they sound subtly different, because the depth of the drum affects the strength of the relationship between the batter and resonant drumheads – it's no surprise that the further away the resonant drumhead is from the batter, the weaker the coupling between the two is. As a result, deeper drums generate less overtones from the resonant drumhead, and therefore often sound to be a bit darker and more weighted towards the fundamental pitch of the drum, which can be great for rock drumming. Deeper drums also tend to project sound more (i.e. they are louder), because the physical mass and energy of the drum and the air inside is greater than with a shallower drum. The overbearing principle is that the depth of the drum predominantly affects the timbre of the sound (remember "timbre" from Chapter 8!) in the same way that the shell material does too. So, whilst it's an important factor for the overall sound of the drum, it's not actually a significant measurement with respect to tuning the pitch of the drum.

9.3 Pitches and intervals on the kit

Before deciding exactly what size drums and what types of drumheads you will prefer to use, it's valuable to think of the type of drum tuning you are aiming for with the drum kit. If you know what sound and tuning you are aiming for, then you can make good choices when selecting a drum kit or when buying new drumheads in order to achieve that sound. Of course, if you already have a drum kit (and hence cannot change the size of your drums!) and/or are not in a position to buy a full set of new drumheads, then it's also valuable to know exactly how to get the best out of your kit and what options are available with your current set-up.

Regardless of how many drums are in your drum kit, tuning is a valuable process to get the best out of each drum. If you have lots of drums in your kit, then they should all give a uniquely different sound; otherwise, what is the point in it being there? So you can only really justify having six rack toms if you are going to tune them all to sound different. This might seem obvious, but it is very possible to tune two drums of different sizes to exactly the same fundamental frequency or pitch, and, although they will sound subtly different, this really isn't making the best use of having two different sized toms in your kit. Equally, if you have a simple four-piece kit and just one floor tom and one rack tom, then it is perhaps even more important that each drum is tuned to a very well-chosen pitch and, hence, give you a musical range for the full kit which can be exploited in a performance.

The key to tuning with a full drum kit is to consider the musical intervals between each drum; this means the relationship between the different pitches of each drum in your kit. Even if you prefer to not tune your drums to musical frequencies, the difference between each drum's fundamental frequency is influential in making a drum kit sound expressive and exciting. As a very simple example, we might tune all of the drums in a five-piece drum kit to have fundamental frequencies at 25-Hz intervals, such as the following:

- 20" Kick drum = 75 Hz
- 14" Floor tom = 100 Hz
- 12" Rack tom = 125 Hz
- 10" Rack tom = 150 Hz
- 14" Snare drum = 175 Hz

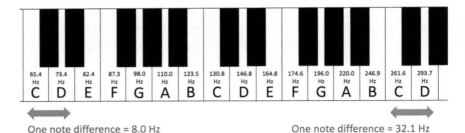

Figure 9.3 The difference in frequency between one note at low and higher octaves on the keyboard.

Generally, the kick drum is the lowest tuned drum in the kit, and the snare is the highest, though these are not strict rules that have to be followed. We know that the drum diameter is related to the drum's fundamental pitch, so the larger toms are tuned to lower frequencies than the smaller toms.

Actually, the above example isn't quite ideal for most cases, because leaving an equal frequency interval between each drum doesn't conform to the most basic principles of musical intervals, which are based on a logarithmic scale rather than a linear one. What this means is that the frequency difference between two musical notes is greater at higher pitches than it is at lower pitches. If you look back to the musical frequency chart we showed in Chapter 3, or the one here in Figure 9.3, you'll see that the frequency difference between adjacent C and D notes on the left-hand side of the piano is just 8 Hz, whereas the difference between C and D notes on the right-hand side of the piano is 32.1 Hz. That's quite a big difference; in fact, a range of 32 Hz on the left-hand side of the piano covers no less than eight musical notes from C2 (65.4 Hz) to almost G2 (98.0 Hz), yet the same amount covers just one note higher up the piano.

Given that we can measure the fundamental pitch or frequency of a drum, we can identify exactly where each drum's pitch lies in relation to a piano keyboard, and this helps visualise the potential range of drum tuning and the intervals between drums. For example, depending a little on what drumheads are used, a 16″ floor tom can usually be tuned between around D2 (73.4 Hz) and G2 (98 Hz), and a 14″ floor tom can be tuned similarly but shifted up a note or two, so from E2 (82.4 Hz) to A2 (110 Hz), as shown in Figure 9.4.

We can see there is a considerable overlap here, so again this allows flexibility for the drummer to choose whether they want a large diameter drum tuned tight or a smaller diameter drum tuned looser, if they are aiming for tuning to the notes E2, F2, and G2, which overlap in the possible ranges for the 16″ and 14″ drums. The overlap of tuning ranges applies similarly for many drums; for example, many kits are supplied with three rack toms of 12″, 10″, and 8″ diameter. A feasible range of tuning for these drums is shown similarly in Figure 9.5.

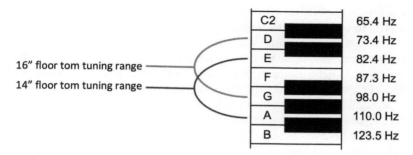

Figure 9.4 Example tuning range for 16″ and 14″ floor toms.

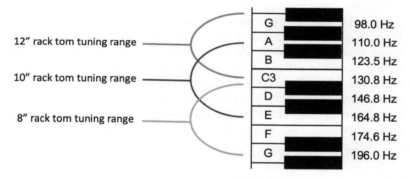

Figure 9.5 Example tuning range for 12″, 10″, and 8″ rack toms.

Interestingly, from Figure 9.5, if you want to tune two toms to fundamental pitch of A2 and E3, you could feasibly do this with two identical 10″ drums that are tuned differently, or you could achieve the same result with one 12″ and one 8″ tom, or other combinations. At the start of this chapter, we mentioned Terry Bozzio's kit, in which he uses 14 identical piccolo toms to cover a full octave of notes, which evidences that most drums can be successfully tuned to a number of different pitches.

TRY FOR YOURSELF: INTERVALS BETWEEN TOMS

Take two toms that are close but different sizes, ideally either two floor toms (16″ and 14″ perhaps) or two rack toms (13″ and 12″ perhaps), depending on what you have available. Tune them both to have a fundamental frequency somewhere in the middle of their range and to take a

(Continued)

reading of each drum's fundamental. Now tune each drum up or down a little to be close to the nearest musical pitch on the music frequency chart. Play the two drums together and explore some interesting musical patterns, maybe some paradiddle patterns or simple polyrhythms. You should get a sense of musicality from what you are playing and feel like the drums can themselves deliver musical phrases.

Now add a third tom drum and tune this to a different musical pitch too. Again create and play some musical patterns on the drums. You should be able to consider the performance phrasing as similar to an arpeggiated pattern which a pianist might play with the fingers on one hand or a guitarist picking through the strings of a held chord. If you have more drums, start to add these in and see how complex and musical you can get with your tuning and performance.

Another good exercise is to take two closely sized drums, for example a 14″ and 16″ floor tom, and tune them to the same frequency. If you have already tuned two of these drums to a midpoint in their range, then tune the larger drum slightly higher and the smaller drum slightly lower in pitch, and you should be able to get them to meet at the same frequency with a little fine tuning. Now listen to the two drums as you play them. They have the same pitch but some subtle differences in timbre because of the different drum sizes and drumhead tension. Which drum do you prefer the sound of and why?

9.4 Tuning suggestions for different genres

It's generally accepted that rock and metal drummers tend to tune to lower frequencies, and jazz and funk drummers tend to tune to higher frequencies, with pop and fusion drummers tuning somewhere in between. Obviously, this is a big generalisation, but a very good starting point to consider when experimenting with your own drum tuning preferences. Equally, it is possible to tune the fundamental pitch of tom drums to intervals that are musically related. While this is not essential, it is an interesting concept which many drummers explore. As an example, if you tune your floor tom to an E2 note (82.4 Hz), you may then also tune a rack tom to be at a musical fifth of that pitch, which is B2 (123.5 Hz), as shown in Figure 9.6a. This can work really well, but if you add a third or fourth tom to the kit, then it's not easily possible to maintain a relationship of musical fifths, because the frequencies become too far spaced apart for the drums to be tuned to. When adding more toms, it's generally wiser to reduce the intervals so that a similar or just slightly larger frequency range can be covered, with more pitches (from more drums) covering the range from lowest to highest. For example, a set-up with two floor toms and two rack toms might start at a slightly lower frequency (say D2 = 73.4 Hz) and use intervals of major thirds. So the

(a)

(b)

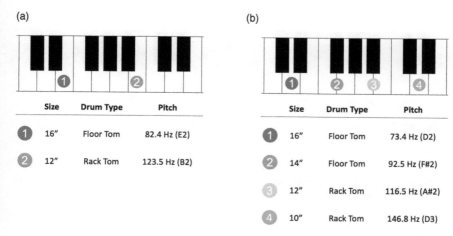

	Size	Drum Type	Pitch
①	16"	Floor Tom	82.4 Hz (E2)
②	12"	Rack Tom	123.5 Hz (B2)

	Size	Drum Type	Pitch
①	16"	Floor Tom	73.4 Hz (D2)
②	14"	Floor Tom	92.5 Hz (F#2)
③	12"	Rack Tom	116.5 Hz (A#2)
④	10"	Rack Tom	146.8 Hz (D3)

Figure 9.6 Suggested fundamental frequency tuning for toms with (a) two toms tuned to an interval of musical fifths and (b) four toms tuned to major third intervals.

second floor tom is tuned to F#2 (92.5 Hz), the first rack tom is tuned to A#2 (116.5 Hz), and the second rack tom is tuned to D3 (146.8 Hz), as shown in Figure 9.6b. If you are not up to speed on your musical scales, these choices might seem a little unobvious at first, but referring back to the major scales interval chart in Table 5.1 gives the frequency multiplying factor, so to tune to intervals of fifths, multiply the lowest frequency by 1.5, and to tune to major thirds the multiplying factor between each drum is 1.26. Alternatively, you may choose to use intervals that are more related to the notes or chords that are predominant in the song being performed, which might be in a major, minor, or other musical key.

While the above examples make musical sense, it's often necessary to adapt the tuning approach for different situations, with different drumheads and a different number of drums in the kit. It's often best therefore to have a slightly different strategy than one related directly to musical scales; first, decide what you want the performance range of your drums to be, and then decide what pitches to tune the drums to cover the frequency range in between the highest and the lowest of the kit. Indeed, as discussed before, it is not essential to tune your drums to musical frequencies, though aligning examples and suggestions with musical pitches allows the discussion to be more connected to the realms and languages of music, which is valuable when discussing intervals and frequency ranges for example.

With all this in mind, it's possible to make some suggestions for drummers who would like to aim for a particular sound in their tuning or learn about the range of sounds that can be achieved with their own kit and set-up. Figures 9.7–9.9 therefore show some example low and high tunings for drum kits with two, three, and four toms.

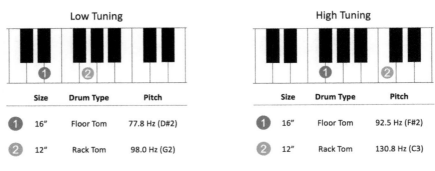

Figure 9.7 Example high and low tunings for a drum kit with one floor tom and one rack tom.

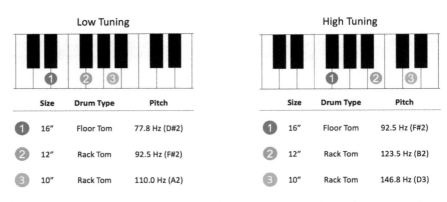

Figure 9.8 Example high and low tunings for a drum kit with one floor tom and two rack toms.

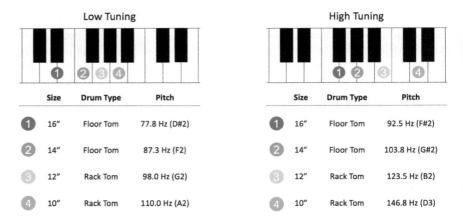

Figure 9.9 Example high and low tunings for a drum kit with two floor toms and two rack toms.

The snare drum is an interesting case, because there are many adopted tuning ranges for the snare regardless of the style or genre being played. Drummers playing rock and jazz (and all other genres) are equally likely to prefer a low or high tuning for the snare. Tuning the snare low gives a warm powerful sound, and tuning the snare high gives a sharp, cutting sound, and the low and high snare tunings are equally valid for all genres. The only difference is that many jazz drummers prefer a 13″-diameter snare (instead of a more common 14″ snare) which allows a slightly higher frequency range and to get just a little higher overtone to the sound. So regardless of what genre you play, it's worth experimenting with low and high tunings on the snare to find your preferred pitch, and maybe you'll adjust this for different songs too. You'll find that with most common drumheads, you can explore the following fundamental pitch range for a snare drum:

- 14″ snares: 164.8 Hz (E3) – 220.0 Hz (A3)
- 13″ snares: 174.6 Hz (F3) – 233.1 Hz (A#3)
- 12″ snares: 196.0 Hz (G3) – 246.9 Hz (B3)

A quick word of warning: some snare heads are heavily coated and damped, which means they are unable to tune as high as lighter uncoated or single-ply drumheads, so be careful when tuning a snare to its upper limit. You may find that the drumhead cannot quite get to the highest frequency mentioned above before it breaks – so take good care and if the drumhead feels like it cannot go any tighter, then don't try to push it beyond its limit! Remember, the drum's overall pitch is affected by the tension of both drumheads, so if you are looking to get a very high pitch of a drum, you'll need to tighten both the batter and resonant drumheads up towards their limit.

TRY FOR YOURSELF: FULL-KIT TUNING RANGE

Set up your drum kit with two toms and try first to tune the fundamental pitch of each drum to those in the low tuning set-up shown in Figure 9.7. Play for a short while and get a sense of the range of the drums in the kit.

Now tune the rack tom up to the pitch shown in the high tuning example in Figure 9.7, if your drumheads allow (remember to tighten both the batter and resonant drumheads to raise the drum's pitch evenly). If it feels too tight on the drumheads, stop at a note lower on the musical scale. Now play the drums again - your rack tom will be way higher and a little out of context, but this will show you the range you can cover with tuning. Tune the floor tom up to the higher frequency and play the drums again.

(Continued)

You may find that somewhere in between these two extremes suits your personal preference, or a hybrid of both approaches. The key is to evaluate the sound and the range of options, and then experiment until you find your preferred style and set-up.

Now add another tom or two and repeat the exercise for the tunings in either Figure 9.8 or 9.9. By the end of this experimentation, you should have a good idea of how many drums you would like to have in your kit, and you should have a good idea of where you want to tune each drum within the feasible frequency range too.

9.5 Drumheads for different music genres

The drumheads you choose equally affect the drum sound with relevance to the style of music you are playing. As discussed in Chapter 7, jazz drummers tend to like hearing the natural overtones ring out, whereas rock and metal drummers tend to prefer a deeper sound with more fundamental and damped overtones. Hence, rock and metal drummers often use drumheads with overtone control features to achieve the sound they desire. The influence that the drumhead type has on the actual fundamental pitch of the drum is mostly related to the thickness and density of the drumhead. We saw previously that thicker and heavier drumheads (i.e. double-ply or coated drumheads) cause the drum to vibrate at a lower frequency than equivalent clear or single-ply heads, so if you are aiming to tune at the bottom of the range for your drum size, then a coated or double-ply (or coated double-ply) drumhead will allow you to achieve that without the drumhead becoming too slack to vibrate properly. Equally the opposite is true; if you want to tune to the top end of the range for your drum sizes, then a clear and more lightweight drumhead will allow you to do this without risk of the drumhead breaking as you tune it up towards your target. With this in mind, drumhead manufacturers tend to advise what types of drumheads suit which types of music, but of course, as we discussed earlier, you are more than welcome to break all the rules in search of your own drum sound and tuning.

Furthermore, the drumhead type can help differentiate a little more between sub-genres. For example, we've seen that metal and rock drummers tend to use quite similar drum kit sizes and quite similar low tuning approaches. But the drumhead choice can make the subtle difference that turns a rock set-up into a metal set-up. For example, a thick coated drumhead sounds great for rock and metal, but the coating gives a much sharper and crisp attack to the sound, which suits metal drumming particularly well. A rock drummer may therefore use the same kit and similar tuning for their kit as a metal drummer, though using a clear double-ply overtone-controlled drumhead to get the deep powerful tone, but with a slightly softer attack to the sound that a metal drummer might prefer. So, despite all the

Table 9.1 Drumhead types and suggested music genres

Ply	Coated	Overtone Control	Musical Styles / Genres
Single	–	–	pop / funk / jazz
Single	Yes	–	pop / R&B / funk
Single	–	Yes	rock / pop / R&B
Single	Yes	Yes	rock / indie / pop
Double	–	–	pop / R&B / funk
Double	Yes	–	rock / indie / pop / R&B
Double	–	Yes	rock / indie / pop
Double	Yes	Yes	metal / rock

permutations and options, the general rule of thumb is a good starting point for your own experimentation. Table 9.1 shows a selection of different drumhead types and the suggested genres which you might use them for.

That's pretty much everything you need to know to put together the perfect kit, set-up, and tuning for the style and sound you are looking for. If you're particularly interested in exploring tuning and kit set-up for rock, indie, and metal drums, there is an excellent overview and detailed discussion in Mark Mynett's book *Metal Music Manual*.[4] Despite being mostly focused on metal music production, the discussions related to drums in this book are relevant to any music genre which requires a deep and impactful drum sound. Of course, everything related to the drum sizes, the drumhead choice, and the tuning range all interacts with each other, so every choice narrows down the set of future choices you might make. It's a fun journey and one that's worth going on, because as soon as you experience how all these things relate in reality, on your own kit and heard with your own ears, you are well on the way to being a master of drum sound!

Notes

1 *The Big Kit* described by Terry Bozio with 360-degree video and musical examples available online at https://terrybozzio.com/about-terry/kit-setup/the-big-kit/ [accessed 01/08/2020].
2 *10 Ways to Sound Like John Bonham* by John Natelli, *Drum! Magazine*, January 2013, available online at https://drummagazine.com/10-ways-to-sound-like-john-bonham [accessed 01/08/2020].
3 Interview with drummer and music producer Emre Ramazanoglu conducted on 14/10/2020.
4 *Metal Music Manual* by Mark Mynett, 2017, Focal Press, pp. 35–49. This book also includes some very relevant discussion on drum recording and mixing too, which is applicable for all music genres.

10 Snare drum tuning

Everyone loves a superb snare sound, even those who don't know it! Listeners involuntarily tap and clap along to the backbeat, and if it cuts through a track with power and style, then it can make the hairs on your neck stand up. In fact, drummers and music producers are generally very particular about snare sounds, both in live performances and when recording or mixing a song in the studio. It can be one of the true defining *sonic signatures* of a recording, and we often crave the "perfect snare sound" in the same way that a guitarist can spend years developing their ideal tone combination of guitar, effects pedals and amplifiers. Similarly, a pianist might be excited to play one type of piano but feel another, however expensive, just doesn't fit their style. To drummers, we have our snares and we are very passionate about them!

Drummer Alex Reeves perfectly sums up the charm and significance of snare drums:

> The snare drum is such an important instrument in pop music – I think that the snare sound, more than any other instrument on the drum kit, sets each drummer and producer apart. When you hit it does it go "blat", does it go "pop", "ping", "clang"? It's all affected by drum choice, head choice, how you hit the thing, what mics and where they go etc. So many choices! I'll often use the same main drum kit throughout an album recording but will almost always change snares between songs.[1]

10.1 Key aspects of snare tuning

Of course, there isn't a single perfect snare, and we're lucky that there are so many different types out there to experiment with – made of solid materials, heavy wood, ply constructions, steel, bronze, titanium, even clear acrylic. What's more, we can put different types of drumheads on to change the sound characteristics and tighten the snare wires to our taste, and even when we have our perfect combination, every snare can be tuned in many different ways. So how do we make sense of finding the perfect snare sound, and how do we go about finding it with our very own snare drums? It's first worth looking back at the most fundamental principles with snare drums; there are perhaps three most valuable things to consider when setting up and tuning a snare:

1   To ensure that the drumheads are tuned evenly and accurately
2   Choosing the drum, drumheads and snare wire set-up to achieve your preferred timbre
3   To optimise the drum to a tuning and timbre that is to your personal preference

When considering the first point – of evenness and accuracy – like all drums, we usually want the drumheads to be evenly tuned so that they vibrate with a smooth vibration profile, with uniform overtones at each lug position, and no beat frequencies such as those discussed earlier in Chapter 4. Equally, we want the two drumheads to be tuned well relative to each other, as discussed in Chapter 5 previously; we want the two drumheads to be operating together and creating a strong, rich tone for the drum, which, of course, is also dependent on the types of drumheads used on the batter and resonant sides.

Choosing the set-up of the drum and tuning the pitch to personal preference is the fun bit. All snare drums sound great at a few different frequencies, tuned low and powerful or tuned high and cutting; it really depends on your personal taste and the type of music you are playing. So it's worth tuning a snare to two or three different pitches and deciding for yourself which suits your style and music. You will probably also find that different snare pitches work better on different songs. Here's a quick pointer for starters: if using a standard 14″ snare, you'll most likely find it sounds great tuned low with a fundamental pitch of about 160 Hz and tuned high at a pitch of about 200 Hz, and also anywhere in between. Any lower than 160 Hz on a 14″ snare is usually a bit flat and boomy, and pushing much higher than 200 Hz starts to get to the point where the drumheads are so tight, they become choked and may even break. Of course, thinner drumheads and smaller diameter snares can go up to higher frequencies, so if you are using a 12″ or 13″ snare, then you'll need to experiment to find the drum's full tuning range.

10.2 Holistic approach to snare tuning

The snare is the drum that requires its heads changing perhaps most regularly, owing to its relatively high tension tuning and its starring role in popular music performance. A holistic and systematic approach can be followed and perfected in tuning the snare from scratch, as described in this 12-step process:

1   Start with all lugs loose on both batter and resonant drumheads.
2   Tighten the batter head tension rods all to finger tight, to ensure that each lug has a similar tightness – even though this will still be too loose for the drumhead to vibrate properly

3   Give each lug a 90- or 180-degree turn with the drum key, using a star form approach; this brings the drumhead up to a sensible tightness so it can vibrate.

4   Perform steps 2 and 3 on the resonant drumhead also.

5   With the snare wires disengaged, listen to the pitch of the drum, or take a frequency reading with the iDrumTune app at the centre of the drum. The pitch will no doubt be too low for a really good snare sound, but it's valuable to have an understanding of what pitch the drum gives with just a simple assembly.

6   Aim to tune the drum up to your preferred pitch. At this point it makes sense to perform the same action to both the batter and resonant drumheads, to ensure they stay fairly evenly tuned relative to each other. To do this, give each tension rod on the batter head a quarter turn (90 degrees) at a time, and then do the same on the resonant drumhead, again following the star pattern.

7   With the snare wires off still, listen to the pitch of the drum and take a reading at the centre of the drum with iDrumTune. It should have gone up from the previous reading and the drum should start to sound quite good. Put the snare wires on and see how you feel about the overall sound.

8   Double-check the evenness of the batter head with the snare wires off, by listening to the drumhead's overtone at all lug locations around the drumhead. The iDrumTune Lug Tuning feature works well for this task. Go around the drum and make sure all the lug positions are within 1 or 2 Hz of each other by making a few adjustments where necessary.

9   You might want to experiment taking the drum to a higher frequency. So give each tension rod another quarter turn, again using a star pattern. Do this on both the batter and resonant drumheads.

10   Now assess the drum's fundamental pitch again – it will be higher in frequency than the previous set-up. Maybe you prefer the drum at this pitch, or maybe you want to try to take it higher still. You need to keep going up in small increments, tuning up both the batter and resonant heads by similar amounts. Be careful to identify when the drumhead is getting to its limit, you don't want to tighten it to the point where it breaks!

11   Once you have the drum at a fundamental pitch that you really like, set the snare wires off and again perform lug tuning to make sure the drumhead is evenly tuned. It's generally less important to perform lug tuning to such a high degree of accuracy on the resonant drumhead, but you can do this too.

12   Finally, it's worth checking the resonant tuning factor (RTF) of the drum by using the Resonant Tuning mode in iDrumTune. The relationship between the two drumheads should be in a good place if you have followed the previous steps. An RTF value of 1.5 works well, as described in Chapter 5, though some drums sound great with RTF at about 1.6 or a little higher.

The reason we perform the tuning analysis with the snare wires disengaged is because it allows the drumheads to vibrate more freely and, for both the iDrumTune Pro app and our ears, to get a good understanding of the tone of the drum. That way we can make good accurate judgements on how evenly the drumheads are tuned and make a good judgement of the overall pitch from day to day and between sessions. Of course, make your final judgement on your preferred pitch to tune to based on the sound with the wires switched on.

TRY FOR YOURSELF: SNARE TUNING FROM SCRATCH

Try tuning your snare from completely loose tuning rods as described in the 12-step approach, and explore the range of the drum as you tune the drumheads gradually tighter and tighter. Implement the Lug Tuning and Resonant Head Tuning actions as described in previous chapters, and you should be able to achieve a powerful, clear, and musical snare sound.

You should be very comfortable with changing snare heads and re-tuning from scratch regularly. If you are worried that you won't be able to consistently achieve a great sound in your drum, then take readings of the fundamental (centre), overtone (edge) frequencies, and the RTF value, and you will be able to recreate the exact same snare sound over and over again.

With all of your favourite snares, it's good to get to know the different tuning options available to you. Snares sound good at many different pitches, so it's worth taking all of your snare drums through the 12-step process to learn more about their range and capabilities, which in turn will help you decide which snares to use on which particular songs. You can also experiment with different RTF values for the snares and decide what value gives the best balance of fundamental and overtones for your drums.

10.3 Manipulating snare drum timbre

There are many options for manipulating the timbre of a snare drum, given the number of different shell designs available, a multitude of drumhead choices, a very broad tuning range to explore, and the set-up of the snare wires themselves. Cherisse Osei (drummer for Simple Minds, Paloma Faith, and Bryan Ferry) describes her number one timbral quality in a snare drum:

> For me, the most important characteristic of a snare drum is its snap-ability! I like a warm solid sound with just the right amount of buzz, which is also sensitive to dynamics.[2]

Smaller diameter snares obviously allow higher tuning frequencies to be achieved (owing to the drumhead equation discussed in Chapter 7), which can be great for standing out in a mix, though this is often at the expense of power or strength to the drum sound. It's also valuable to note that heavier heads (and this includes double-ply heads) have a lower frequency range than lighter drumheads. So if you want to tune your snare up to higher frequencies, consider using thinner or lighter drumheads on the batter side. Of course, thinner drumheads can be broken more easily, so there is sometimes a practical compromise to be made.

Damping is regularly applied to snare drums as a retrofit solution if the desired snare sound is not achieved. In general, if you are using the right drumheads for your preferred sound, then additional damping should not be required in most situations. Drumhead manufacturers put a lot of design and innovation into creating snare drumheads with built-in damping systems, so if you find you need to add extra damping in the form of o-rings, gel, tape, or other things, then you should consider putting some extra time into experimenting with different drumheads. It's very difficult to add external dampening to a drumhead and not impact the evenness of the drumhead vibration – if you have just spent five minutes ensuring that the drum sounds even at all tuning lug locations, then this good work could be undone by adding some tape or extra mass to a single point of the drumhead.

Snare wires give a very unique sonic element to the snare drum and differentiate the sound from that of any other drum in the drum kit. It's important that the snare wires don't unevenly affect the resonant drumhead, so ensure they are positioned evenly across the diameter of the drum. Different snare systems exist for tensioning the wires onto the drum, and those which allow an even tension across the drumhead to be applied (rather than tight to the head on one side and less tight to the head on the other) generally give the most impactful sound. This is because the snare drumheads are enabled to vibrate evenly and smoothly whilst getting a good characteristic from the snare wires themselves. It's good advice to tension the snare wires so that they are not loose and rattling, but equally not too tightly locked against the resonant head, which chokes the drum and stops the heads vibrating properly in unison. Of course, there are no rules in realty – so if you like the way it sounds, that's great!

Shell material and depth give a substantial influence to the drum's timbre, which can only really be understood by listening. Wooden shells vibrate at lower frequencies than metal shells, so a deep thick oak snare will give a subtle deep warmth to the sound of the drum, whereas a thin bronze shell snare extenuates the high-frequency characteristics of drum timbre. Different shell materials also inherently have different elasticity characteristics, which influences how much a solid material can vibrate when a force is applied. The elasticity characteristics of different woods, metals, and acrylic (and their associated manufacturing methods) all therefore influence subtly different characteristics in the snare timbre. You may also want to explore and research novel and innovative snare designs, which are intended to give unique timbre characteristics to a snare drum. For example, free floating snares, which have minimal hardware mounting and attachment to the drum shell[3] and vented snares, which allow some of the air mass within the drum to escape and alter the acoustic coupling relationship between the two drumheads.[4]

TRY FOR YOURSELF: SNARE DRUM TIMBRE

It's really useful to experiment with all the timbre variations for snare drums: different heads, different depths, different snare wires, and, of course, different snare shell materials. It is valuable to make your observations with all drums tuned to the same frequency each time. So, for example, listen to the drum with different timbre set-ups but always at a common fundamental frequency and RTF value (e.g. F0 = 196 Hz and RTF = 1.5).

There are lots of snare timbre experiments presented by the Sounds Like a Drum YouTube channel, so check some of these out for inspiration and to hear what other unique modifications can influence the timbre of your snare drums.[5]

10.4 Comparing snare timbre example

Measuring the fundamental and overtone frequencies of a drum enables fair comparisons and judgements between the sounds of different drums to be made. It's very hard to compare the sound of one drum to another without being sure they are tuned identically. Only when you can be sure that two drums are tuned to exactly the same pitch and overtones, you can decide if you prefer the sound of one or another, or make a judgement on the qualities of the different drums and set-ups. In an experiment, three 14" snare drums were compared, all of different shell depths and with different shell materials and built using different construction methods (and all with different drumhead types too).[6] The drums were all tuned to exactly the same frequencies at each stage of analysis, enabling a fair judgement on the timbres of each drum.

The three 14" snare drums compared were as follows:

- De Broize Custom, solid oak, 6.5" depth, coated Remo Black Spot drumhead
- Ludwig Supralite, steel, 6.5" depth, coated Ludwig standard snare drumhead
- Tama Superstar, six-ply birch, 5.5" depth, coated Remo Pinstripe drumhead

What's great is that when these drums are tuned well, they all sound fantastic at a number of different pitches. They were all tuned to work well from

160 to 200 Hz, although they all have a subtly different character. Subjectively, the De Broize drum has a very warm timbre, which no doubt comes from the solid wood design, and provides a good balance of tone between depth and sharpness. This implies it will work well for many genres and styles, having a sonic "richness" that many drummers and studio engineers desire. The steel Ludwig unsurprisingly has a cutting character and some detailed overtones that do not become overpowering of the fundamental pitch. This drum is ideal for rock and metal when cranked high but also as a very musical sounding drum when tuned lower, so can suit other styles and genres too. The Tama Superstar is surprisingly reliable for a budget snare and stands up alongside the more expensive drums. It seems to come into its own when tensioned to higher frequencies and the damped Remo Pinstripe drumhead allows the overtones to stay controlled, giving the drum a clear pitch and overall tone that can suit many styles and genres.

In the tuning experiment, the snare drum sounds were recorded with two microphones, an AKG C414 positioned approximately 1 meter from the drum and a Shure Beta 57 positioned approximately 10 cm from the drumhead. An interesting point from the comparison experiment is that, in the room and in the AKG C414 overhead microphone recording, the drums sound really quite similar with just subtle perceivable differences in sonic characteristics. But when the close Shure Beta 57 microphone sound was mixed into the playback, the differences between the drums and their different drumheads become a lot more apparent. This shows the benefit and sonic opportunities when a close microphone is used on a snare during a recording session. Indeed the act of drum recording is regarded as a creative skill in its own right, and it's not uncommon for tuning issues to go somewhat unnoticed until you hear the drum sound captured through a close proximity microphone.

Notes

1 Interview with drummer Alex Reeves conducted on 07/10/2020.
2 Interview with drummer Cherisse Osei conducted on 27/10/2020.
3 Free floating snare drum design, as described by Pearl Drums online at https://www.pearleurope.com/product/free-floating-task-specific/ [accessed 01/08/2020].
4 Variable vented snare drum design with the "Air Control System", as described by Odery online at https://odery.com/custom/air-control/ [accessed 01/08/2020].
5 Snare timbre experiments by Sounds Like a Drum, for example *How Snare Wire Count Affects Sound*, available on YouTube at https://youtu.be/3n1sm1ELQ8A [accessed 01/08/2020].
6 Snare comparison video by iDrumTune, *Comparing De Broize Solid Oak, Ludwig Supralite Steel & Tama Superstar Birch Ply*, available on YouTube at https://youtu.be/d4RycHdt4dk [accessed 01/08/2020].

11 Kick drum tuning

Prior to the kick drum pedal being invented, drummers would predominantly be regarded as the bass drum payer, who would play a large drum with a handheld mallet in military or town concert bands. Local concerts gained popular appeal in the US after the end of the American Civil War in 1865, and by this time the bass drum was regularly played with a cymbal attached too. Soon after, the concept of *double drumming* emerged with the drummer now performing with two or more percussion instruments: bass, snare, and cymbal with two handheld drum sticks or mallets. However, it wasn't until the early 1900s when the bass drum kick pedal started to emerge, first seen from innovative modifications by drummers and then engineered into William Ludwig's patent kick pedal design in 1909.[1] Ludwig's design is not dissimilar to that which we recognise nowadays and, importantly, incorporated the spring-back design that allowed drummers to play the bass drum with more power, consistency, faster, and for longer than previously was possible. The result of this innovation meant drummers could explore different low-frequency sounds than those which had been heard before in music. The drummer could now play unique and intricate phrases with their foot and was encouraged to perform sitting down. Once this paradigm shift was in motion, the vision for drumming using all four limbs became apparent, and the modern drummer was born.

The kick drum isn't just heard, its low-frequency punch can be felt physically in our bodies; the human chest cavity has a natural frequency itself at around 30 Hz, which is close to the sub-frequencies of a low-tuned kick drum. Drummer Cherisse Osei emphasises this quality, which she likes to hear and feel in her kick drum sound:

> I like the sound a kick drum to be tight and punchy, with thick low-end and a powerful snap to cut through the rest of the instruments. It has to be thumpy and to resonate in my chest and bones![2]

Kick drum tuning can be a little daunting for many drummers. In reality, with some simple techniques and a little knowledge on drumhead vibration, kick drum tuning can actually be a very straightforward task that doesn't need to be overcomplicated.

11.1 Kick drum tone and dynamics

The physics of the drum allows three general approaches to tuning a kick drum, and each approach will suit different drummers, different music genres, different drum sizes, and different drumheads. The three approaches all relate to the balance of the impact power and the fundamental tone of the drum, both of which can be controlled by damping that is built into the drumheads or additional damping added to the drum. Considering the kick as an extra-large tom, we know that the size and drumhead choice have a significant impact, and generally there shouldn't be a need to add additional damping. However, if your drumheads don't give the sound you are looking for, a bit of extra damping can often improve the situation. Extra damping can also be useful if you want the drum to have a high and fairly musical pitch, but you don't want every hit with the pedal to ring out for a long duration.

Figure 11.1 shows three simplified kick drum waveforms to help explain what the suggested different approaches to kick tuning are.

Option 1. Low and loose

Low and loose, shown by Figure 11.1a, is a great option for drummers who want a sudden and powerful sound to their kick drum. The low tuning essentially refers to tensioning the head just to the point where the wrinkles are gone, and it vibrates properly, but no further. In the waveform of Figure 11.1a, we see this approach giving a low-frequency peak with no ongoing vibration; the resultant sound is a deep thud that starts and stops very quickly with little overtone. This tuning is generally only possible to achieve with heavy drumheads that have built-in overtone control. Low and loose works great for rock and metal drumming, but you'll find it in all kinds of music where the kick drum sound drives the song significantly.

Option 2. Tight and tuneful

If you tighten the batter and resonant drumheads to a higher tension, the fundamental tone of the drum starts to dominate and gives a more musical sound. Figure 11.1b shows a kick drum waveform with a tighter frequency peak and some ongoing vibration profile of both the fundamental and overtones of the drumheads. This drum sound still has a

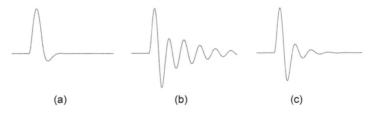

(a)   (b)   (c)

Figure 11.1 Simplified waveform diagrams showing three different kick drum profiles.

powerful impact, but with some character and tone that carries on after the initial hit. It's no surprise that jazz, folk, and more classical drummers tend to use this approach, but also some rock and pop drummers like to be musical with their kick drum and experiment with this approach too.

Option 3. Damped tone (hybrid)

The third option sounds like a hybrid of both options 1 and 2. The drum is tuned higher to give some musical tone, but is heavily damped (by using damped drumheads or some additional damping applied to the drumheads) in order to keep the impact quite sudden and powerful, and to significantly reduce the overtones of the drumheads too, as seen in Figure 11.1c. The sound has a controlled tone that can be tuned to suit the style of music or to give some character without causing big boomy sounds that comprise a performance or recording. It works particularly well for indie, pop, and R&B genres, and has been adopted conceptually for electronic music too, where synthesised kick drum sounds are designed to have both impact and musicality.

11.2 Drumheads for the kick drum

The drumheads used are very important for achieving the kick sound you want, and if you use the right drumheads, it should be possible to get the sound you are looking for without any need for additional damping. As with tom and snare heads, there are a huge range of options with kick drumheads to consider. If you are aiming for the low and loose tuning, then you will want heavy drumheads on both the batter and resonant sides (either coated or double-ply), and you'll probably want some overtone control built into the drumheads too. For the tight and tuneful approach, you can use the same drumheads, just tuned up tighter; however, you may find that at the fundamental frequency you are tuning to, the drum just doesn't sound bright enough. If you are looking for a bright tuneful sound, then use a single-ply clear drumhead on the batter side, and this should give a little more overtone. Again, built-in overtone control may be what you want, even for tight and tuneful, in order to keep the balance of the fundamental and the overtones fairly equal, but you may prefer a slightly softer or more subtle overtone control system to those designed for rock and metal setups.

For the hybrid damped tone approach, you can use almost any drumhead, and so this is a good approach if you have no funds to splash out on a new kick drum batter head. That said, investing in good and suitable drumheads is the single best thing you can do for your drum kit, above anything else. Ideally, you can achieve the damped tone sound without using any damping or muffling other than that which is incorporated into the drumheads; if you use a double-ply or coated batter head with overtone control then you should be able to achieve a tuneful-yet-damped sound. A good approach is to, as with Option 1, tune the batter head to the point where the wrinkles have just gone and the drumhead is just able to vibrate properly, and then go no further. Then, slightly increase the tension of the resonant drumhead to find the tone you are looking for. If you don't have heavy or muffled drumheads, you can get a similar effect with some additional damping in the form of a cloth, blanket or some foam positioned inside the drum and gently resting against one or both of the drumheads. Adding damping inside the drum both obstructs the coupling between the two heads (which dampens the fundamental tone of the drum) and if the muffling is touching the drumhead at the edge then this also acts as an overtone damper. Start by just damping the resonant drumhead with a little contact from a blanket inside the drum, and if the drum still rings out a bit too much then move the blanket to dampen the edge of the batter head too. Some drummers have also experimented with a small towel or cloth wedged behind the kick pedal and the batter head to dampen the overtones, but this is a bit of a last resort solution. It's much better to get the damping you need from the drumheads, if you need to add much additional damping then it implies you are using the wrong drumheads! The use of a cloth or other material inside the drum brings two or three compromises; firstly it obstructs the air movement inside the drum and stops it vibrating as it was designed and intended to; second it adds damping to just one small location of the drumhead (rather than uniformly around the drumhead in the case of dampers built into the drumhead), and thirdly, the sound of the kick drum is different every time you move the drum kit and while you play, because whatever is inside the drum will always have a different type of contact with the drumheads and can potentially move during a performance or in transit.

If you have the right drumheads and you know what sound you are aiming for, how do you get started? Well, as with all drums, it's can be quite simple – just get all the tension rods finger tight and then tune them up incrementally until you achieve the sound you are looking for. Rob Brown has some excellent videos on drum tuning and very much promotes simplicity and non-complicated approaches to tuning.[3] Remember, the fundamental pitch is affected by the tension of both the batter and resonant drumheads, and often with the kick drum it's beneficial to tighten the resonant head first if you are trying to take the overall pitch of the drum higher, this way you get the damped and low tension response of the beater on the batter head, but the resonance and tone from the influence of the resonant head.

TRY FOR YOURSELF: KICK DRUM TUNING

Once you have the drumheads to finger tight, put a weight (or press down with your hand) in the middle of the drum and make a turn or two on each lug until the wrinkles smoothen out at each point. We suggest using a star form tuning, tightening opposite lugs each time rather than sequentially around the perimeter of the drum, because it is valuable to keep the drumhead uniform on all sides as you tighten up from finger tight – this avoids causing the beating effect which we saw previously when discussing Lug Tuning. Once you have the two drumheads just above the point where the wrinkles smoothen out, you can use the iDrumTune Lug Tuning feature to make sure the overtones are all fairly even around the edge. At this point, have a listen to the drum, if it sounds good then stop, you are done! If you are looking for a particular fundamental frequency, then use the iDrumTune Pitch Tuning feature and hit the drum in the centre of the batter head with a mallet. You'll see the fundamental pitch of your drum and will know how much to tune up or down in order to get the pitch you are looking for.

11.3 Kick drum tuning range

In terms of tuning range, you'll see that the fundamental frequency can be anything from around 55 Hz to around 80 Hz for a standard 24″, 22″, or 20″ kick drum, and potentially even higher for a small 18″ or 16″ kick drum that can be found on some jazz club kits. In each case you'll find the overtone works well at an RTF value of 1.4 or 1.6, as with standard tom drums – resulting in overtone frequencies between around 80 and 120 Hz. Because the fundamental is often heavily damped with kick drums, sometimes you'll notice this is both hard to hear as a pure vibration and hard to read with iDrumTune. The most important thing is to ensure the overtones are fairly even with Lug Tuning and if sounds right, then it is right!

The image below shows a 22″ kick drum tuned to a fundamental F0 frequency of 67 Hz with an F1 overtone at 94.5 Hz, and all lug positions tuned to within 1 Hz of each other. You may find that, depending on the drumheads, iDrumTune picks up either the fundamental or the overtone more powerfully, regardless of where you hit the drum. This is for two reasons, firstly because different kick drum heads give very different balance

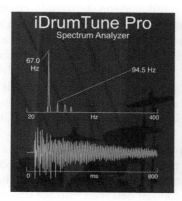

Figure 11.2 Kick drum tuning with iDrumTune.

between fundamental and overtones, but also the low fundamental frequencies are approaching the low limit of a mobile recording device. This is not a problem, it just means that you may need to use iDrumTune's Target Filter mode when tuning a kick drum.

TRY FOR YOURSELF: KICK DRUM TUNING RANGE

Explore the tuning range of your kick drum, to see how low it will go, and how high and resonant you can get it too. A standard 22″ diameter kick drum should tune to a fundamental between 55 and around 80 Hz, depending on what heads you use. If you have damped or coated drumheads, the kick should sound good at all frequencies, even if you have to add a little extra damping as you tune up. When tuning to higher frequencies, first tune the resonant head up and evaluate the sound, then tune the batter head further if you want to try higher still. Experiment with the three tuning techniques – low and loose; tight and tuneful; and damped tone. It's good to be able to dial in different sounds depending on the situation, and an appreciation of the different tuning approaches will allow you to more confidently find the kick drum tuning you are looking for.

11.4 Controlling the kick drum timbre

There are a number of other concepts related to kick drum sound; here are a few other points to mention which you might experiment with:

- To get a more "snappy" or "clicky" attack to the kick sound, use either a coated drumhead or stick on a firm plastic dot where the beater hits the drumhead. A firmer kick pedal beater (plastic or wood) gives a harder attack than a felt beater too.
- The depth of the kick drum has a big influence on the sound – conversely deeper drums do not always sound bigger! The deeper the drum is, the less reflection there is from the resonant drumhead, so the sound energy can get lost a little bit. On the other hand, shallower kick drums can vibrate for longer because of the strong coupling and so you might need more damping if this is too much for your preference. Generally shallower drums are more responsive and you can feel the energy moving inside the drum more with the kick pedal, especially if you are playing a hard and fast pattern.
- We often use a hole in the kick drum's resonant drumhead to let a bit of energy dissipate as an equivalent to drumhead damping, and conveniently to allow a microphone inside the drum when recording. The hole allows energy to escape, but if this is well designed into the drumhead,

it can help to enhance and boost the low frequencies of the drum too by an acoustics phenomenon called the *Helmholtz Effect* (a technique also applied to loudspeaker design). Look it up if you are interested, it's the same acoustics theory that causes a half-full beer bottle to generate sound when you blow over it![4]

- How you play the kick drum makes a big difference to its sound too. If you bury the kick beater hard into the drumhead, this will dampen the low fundamental tone. If you let the beater bounce back off the drumhead, the fundamental tone will ring out more powerfully. Playing softer can often result in a more powerful sound when recording too. This is because with softer hits, the attack or *transient* of the sound is more similar in volume to the *body* or vibrating tone of the drum, allowing a more consistent audio waveform which can be played back at a louder volume – we'll discuss this type of recording and mixing consideration in more detail in the following chapters.

Drummer Alex Reeves explains his approach to kick drum tuning, contextualising many of the concepts discussed in this chapter:

I love a deep, low, dampened bass drum - it can work beautifully for so much music. Personally, for a super radio-friendly sound I'll tune the bass drum low-ish, very little ring, playing the beater off the head so there's more of the low-end tone for the mics to capture, matching the type of beater to the amount of attack we need - wooden beater for more of the front end click, felt beater for a bit more warmth and softness, lambs-wool beater for a super-moody, boomy sound.[5]

Whatever kick drum sound you are looking for, it's worth experimenting with the approaches outlined above, to understand the influence you can have over your low-end sound and to explore the sound that suits your personal style. Try using a mallet when tuning rather than trying to work with the kick pedal, a good drum mat is also recommended, not only to protect the floor, but to add a bit of damping and to reduce the vibration coupling with the floor too. Every little helps when you are looking for the perfect sound!

Notes

1 As discussed in *Kick It: A Social History of the Drum Kit* by Matt Brennan, Oxford University Press, 2020, pp. 31–47.
2 Interview with drummer Cherisse Osei conducted on 27/10/2020.
3 YouTube tutorials on kick drum tuning by Rob Brown, for example *Tune Your Bass Drum Quick & Easy* at https://youtu.be/FTdXOWKIQc4 [accessed 01/08/2020].
4 The KickPort innovation is designed around the Helmholtz Resonator principle, as detailed at https://kickport.com [accessed 01/08/2020].
5 Interview with drummer Alex Reeves conducted on 07/10/2020.

12 Production and preparation for drum recording

Having learnt about the concepts of drum acoustics, drum tuning, and the accuracy and psychoacoustics of listening to drums, all drummers are well equipped to develop their own preferred sound for live performances in a number of genres and scenarios. However, the process of recording drums to be played back for others to enjoy is a creative and scientific field in its own right, and one which drummers and drum sound play a very significant role in. Recording and producing drums in the studio takes the analysis and manipulation of drum sound to an even higher level of detail. As a result, it's not uncommon for drummers to hold the key to the success of a recording project, with respect to setting the timing foundation that recorded songs all build on and the tonal design of the drum sound which can define not only the genre of the song but also the power, intensity, and physical connection which listeners have with the music. Recording and producing drums is a multifaceted task, and drummers benefit by understanding this process and being well prepared for drum recording sessions. It's no surprise that many drummers go on to become great studio recording engineers, mixers, and producers, because their appreciation of timing, groove, and low-frequency sound is so fundamental to a successful recording project.

12.1 Production and pre-production

Great recordings don't just happen, they are produced! Every incredible drum performance captured on record for the world to listen to was not achieved by a single piece of good luck, or purely an inspired performance, but more likely a well-formalised plan that was perfectly put into action, perhaps with a little unexpected magic occurring along the way. The general fact is, unless recording projects are planned, prepared, and implemented to a fine level of detail, they can easily fail and give poor results. Most experienced studio engineers will be able to give examples of unexpected challenges encountered during a recording project, owing to one or more oversights. As a result, many recording projects do not even make it to the mixing stage, because they don't meet the quality expectations when listening back after the recording session is complete. The responsibility of a music producer is to develop a strategy for the entire recording project and to ensure that everything required to achieve the expected sound quality and style in the final artefact is in place.[1] A good music producer will also be agile and be able to modify or alter the plan if they see issues or opportunities arising during the production process. Production for drums therefore involves tying together everything that happens from the moment someone decides a recording is to be made to the moment it is delivered as a final version that needs no further modification. We use the term *pre-production* to refer to all the aspects of a recording project that occur before the day of the recording session (which is sometimes referred to as a *tracking* session). Pre-production can involve many aspects that are *creative, technical,* and *managerial*; indeed the project management aspects of a recording project are not insignificant and if overlooked can be the difference between success and failure.[2] Since time spent in a recording studio is generally a costly component of a recording project, pre-production is a valuable process to ensure that costs are kept down and within budget, and avoid unnecessary experimentation and debate in the recording studio itself.

Examples of creative production and pre-production tasks for a drum recording project include the following:

- Finalising the song composition and rehearsal with musicians
- Choosing the drum kit and its particular setup for recording
- Deciding the performance approach, i.e. whether to record all instruments at once, or making overdub recordings of each instrument individually, with backing tracks to play along to
- Deciding the temporal strategy – will a metronome at a specified tempo be used or will the band play free form to their own time?

Examples of technical production and pre-production tasks for a drum recording project include the following:

- Deciding the most suitable recording studio and performance space
- Microphone choices and placement techniques
- Deciding the signal chain for recording, equipment used, and approach for headphone monitoring
- The mixdown approach – who will do the mixing, when, what types of audio tracks do they need, will mixing be done after the tracking is complete, or will it start during the tracking process?

Examples of project management production and pre-production tasks for a drum recording project include the following:

- Building the right team for the entire recording project and communicating with the team about objectives and individual responsibilities
- Identifying a recording venue, studio booking, and equipment hire

- Time planning and scheduling work
- Financial planning, to ensure the recording project completes without exceeding a specified budget

---

TRY FOR YOURSELF: PRE-PRODUCTION

Before starting your next studio recording project, consider all the creative, technical, and project management tasks that will enable the project to run smoothly. Document these in each category to monitor and check that you have completed the pre-production tasks before the recording starts.

When in pre-production, it's also worthwhile to consider and list all the things that could potentially go wrong during the recording sessions, and make plans as to what you would do in the case of any issues arising.

12.2 Setting standards and getting results

It's valuable to set clear objectives and quality standards for any recording project: What attributes will make the recording sound great? What defines a good drum sound for the genre? How can you achieve the sound you are looking for? In many respects, with a great recording, many subtle aspects defining the drums can be manipulated during mixing, for example, to add greater impact or the right amount of ambience or to decide on the relative volume of the drums with respect to other instruments in the song. However, a good quality recording keeps the door open to many more options during mixing, and often requires less attention or manipulation during mixing too. A good analogy is golf – the best way to improve your putting statistics is to land your pitch shot closer to the hole![3]

Recording is a difficult task that inherently holds many sonic compromises, but if care and attention is paid to each component or link in the recording chain, then great results can be achieved. It's always the weakest link that lets down a creative project with multiple components, so some simple and clear attention to detail is required to ensure the results are to the correct standard and expectation. *The Good Rule* is therefore a very simple but effective methodology to follow for recording, summarised in Figure 12.1.[4]

The Good Rule sounds quite simplistic, but get all those things right, and you can't fail to make a good recording. Of course, you need to connect the mic to a reasonable quality recording set-up and press record at the right time too! If you are acting as a music producer, then it is your responsibility to ensure that all aspects of the Good Rule are followed during the recording process. If any of the elements of the Good Rule are compromised, then

The Good Rule!

good musician + **good performance** + **good acoustics**

+ **good mic** + **good mic placement**

=

good sound

Figure 12.1 The Good Rule for recording music, by David Miles Huber and Robert E. Runstein.

very possibly the whole recording is compromised. The curse of the weakest link can make its way down the rest of the signal chain; even if all the elements of the Good Rule are adhered to, a noisy mixing desk will still give a noisy recording, regardless of how expensive the microphones are. Some aspects are easier to ensure than others; for example, you may be working with musicians who are not performing well on that particular day, or you may not be able to afford a room with perfect acoustics for your project, or you may not have access to the microphones you want to use. In each case, the producer's responsibility is to minimise the weak links in the chain. While you might have help from a sound engineer in setting up the room or the microphones, the influence of the producer in helping the musician to perform to the best of their abilities should not be underestimated. Different musicians and artists respond to different things; some are totally frozen by fear in the recording studio, others become exited and find it difficult to focus, and some are self-conscious and need to feel the right environment before their best performances can come out.

It's also valuable to think towards the mix and the final finished product when recording. If you want the mix to have tight hard-panned, impactful drum sounds, then you need to record with close microphones and use the right drumheads for that sound. If you decide this after the recording has taken place, then there will be limitations on the mix and how many of these poor decisions at the recording stage can be fixed. In fact, the Good Rule describes all the things that are very difficult to "fix in the mix". For example, it is not possible to make a poor singer sound good, even with autotune. It is not possible to make a dead microphone sound alive. You cannot remove unwanted reverb at mixdown. You can't perfectly edit an out-of-time performance to be in time, without bringing other compromises to the mix. You can't boost frequencies that aren't there in the recording. So mixing really does start with recording, and even before that too in pre-production. The better prepared you are from the start, the more likely you are to achieve your objectives and expectations from a recording project.

12.3 Choosing and evaluating the recording space

The room used for drum recording has a huge influence on the sound of the recording captured into the microphones and heard on playback. Microphones over the top of the drum kit (*overheads*), and other *room microphones* capture a significant proportion of sound reflected from the walls, floor, and ceiling, but even *close microphones* placed near individual drums and cymbals can be affected by the room reflections too in some cases.

There are a number of things to consider when choosing a recording space for drums. There are many occasions where the perfect or most preferred room cannot be secured, owing to financial or availability limitations, so a good recording engineer will understand how to get the best out of the room that is available to them. In evaluating the space used for a drum recording, you should consider the size of the room, the materials used in constructing the room (and hence the resultant reverberation characteristics in the room), sound isolation and background noise levels, the suitability for hosting a music recording project, and the positioning of the drum kit within the room should be considered too.

Producer Tommaso Colliva and drummer Alex Reeves conducted a very interesting drum recording experiment, documented in *Sound on Sound Magazine* titled *Recording Drums: What Difference Does the Room Make?*[5] In their experiment, Tommaso and Alex recorded the same drum kit in some of the UK's best recording studios and compared results, identifying the factors which created variations in the different recordings and resulting in some valuable new ideas and insight for both. There are also some excellent and detailed books available on recording studio design and room acoustics for making music[6] – here we'll just take a closer look at each of the above described aspects relating to drum recording, which all relate to each other too.

12.3.1 Room size and dimensions

The size of the recording room has a huge influence on drum recordings. In general, larger rooms work best for drums for a number of reasons. Firstly, larger rooms mean that any reflections to the microphones from the walls or ceiling will have travelled quite a long distance. This means that they will be at a much reduced volume than the direct sound from the drums, owing to the *inverse square law*, which states that the energy of sound reduces exponentially over distance.[7] So doubling the distance which the sound has to travel results in a four times reduction in the energy of the reflected sound. Secondly, the longer distance to the walls or ceiling in a large room means a longer time passes before the sound reflections arrive back at the microphone. This has a valuable effect on the recording by providing a very clear differentiation in the sound that is directly from the drums and that which is reflected from the walls, and the sonic effect of this are much more preferable, at least to most studio engineers. In smaller rooms, the first reflections of the sound (referred to as *early reflections*) can return to the microphones before the actual direct sound of the instrument has completed, which causes significant alteration of the sound and *comb filtering* (a concept which is discussed more in Chapter 14), and which we often refer to as sounding "boxy". This applies to the recording of all instruments and particularly to the human voice when recording someone singing or speaking. For drums, reflections and comb filtering have the effect of stopping the drums from sounding clear and crisp at the moment they are hit and in the projection of the sound. For this reason, many producers and studio engineers attempt to minimise the level of early reflections in a recording space, and many mix engineers reduce the early reflection parameters on artificial reverb plug-ins for a similar reason too. Often rooms suffer from early reflections by having a low ceiling; even if the room appears large enough in its width and length, the early reflections from a close ceiling can still have a significant negative effect on the recorded drum sound. Figure 12.2, for example, shows the direct and early reflections for a drum kit positioned in the corner of a room. The significance of the detrimental short-path early reflections (Figure 12.2b) can be reduced by either moving the drum kit to the centre of the room or using a larger room. Figure 12.2 also highlights more *diffuse* reverberation sounds (Figure 12.2d), which generally cause less

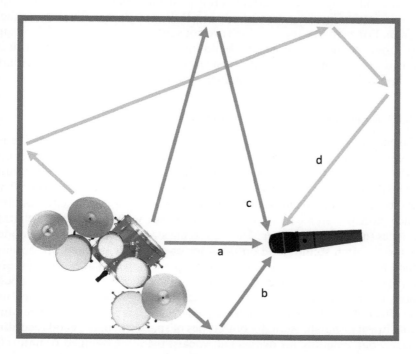

Figure 12.2 Sound transmission from source to microphone: (a) direct; (b) short-path early reflections; (c) long-path early reflections, and (d) diffuse reverberation.

significant reflection issues on the recorded sound, owing to the longer time taken for these sounds to arrive at the microphone and their energy attenuation over the longer distance travelled too. The more diffuse reflections do, however, add a unique (and often pleasant) reverberation characteristic depending on the size and shape of the room, as well as the materials used in construction of the room and the acoustic characteristics of other items positioned in the room too.

A further reason for using large recording rooms is that all rooms have their own natural frequencies, which we call *room modes*. At these frequencies, the wavelength of the sound is the perfect size for the room dimensions and causes a build-up or a peak at the room mode frequencies in any recording. This, along with the reverberation properties of the space, is often referred to as the sonic *character* of the room. But if the room mode frequency somehow aligns with a dominant frequency of one of the instruments or the drum kit, then it can cause a detrimental imbalance to the frequencies captured in the recording. Room modes and frequency peaks of smaller rooms tend to be at frequencies well inside the human hearing range and frequencies which coincide with those of many conventional musical instruments. However, larger rooms have much lower frequency room modes and can be often down at frequencies below those that we can hear or try to record, so larger rooms often have a flatter overall frequency response. This can be improved further by using rooms that are not symmetrical or simplistic in their design. For example, rooms which are not square or rectangular do not support room modes that normally build up between parallel facing surfaces, so this has become a valuable consideration in studio design, particularly for improving the acoustics of smaller rooms.

12.3.2 Room materials and reverberation characteristics

The choice of materials used to build the room has an influence on the recorded sound too. If the room is small and suffering from room modes and early reflections, then often the only solution is to fit the room with substantial *sound dampers* or *acoustic absorbers* to stop the reflections altogether. This results in a slightly dead sound with not much charm or character, but it is much more preferable to a small room recording with reverberation characteristics that cannot be altered at the mixing stage. In larger rooms, wooden materials often sound good on the floor or as panels on the wall, because wood has the effect of absorbing some frequencies whilst allowing a natural warm sound in those reflected back to the microphones. Consider a wooden room in comparison with a glass, tiled or stone-walled space which has the effect of amplifying the high frequencies in the reflection and sounding quite brash or harsh. Actually both types of space can make good recording spaces, but generally for drums and vocals, the more damped or naturally reverberant sound is usually preferred. The shapes of acoustic absorbers can also have a positive effect. Some acoustics control systems, called *diffusers*, are designed to reflect the energy in different directions and cause scattering of the sound, to avoid the build-up of room modes or echo. We also see that absorbers situated a few centimetres away from the wall or ceiling cause an improved absorption effect too, because the sound energy is greater in the room itself rather than exactly at the position of the wall. Some smaller rooms may also use *bass traps* in the corners, which are dense and deep acoustic foam or fibreglass systems that can more effectively tame the large wavelength frequencies of unwanted bass room modes – sofas, armchairs, and soft furnishings can have a similar positive effect too on room acoustics, though to a much less quantifiable degree.

12.3.3 *Sound isolation and background noise levels*

Sound isolation is important to stop a drum performance from annoying the neighbours, but as perfectionist recording engineers we are generally more concerned about the sound coming into the room from outside! Luckily drums are so loud that we don't need to worry as much about a gentle amount of hum from a nearby road or the occasional aeroplane going overhead – these things can cause big headaches when recording quieter sounds such as vocals or speech. But, there may be issues in the room itself, such as a buzzing strip light or an air-conditioning unit that will add an unwanted hiss or hum to the recordings, which can particularly get picked up by overhead and room microphones. These noises should be avoided where possible; it's not uncommon for sound engineers to carry around a bunch of socket-powered table lamps when recording in an unconventional room, because they guarantee no unwanted noise generation and provide a nice creative ambience too. You may even find some engineers prefer to switch the air conditioning off and work in a warm room – there's clearly a trade-off here and you may find that some rooms, particularly those in colleges or town halls for example, do not allow access to the air conditioning controls. For this reason, if you are not tracking in a conventional recording studio space, it's worth checking out the room beforehand as part of the pre-production tasks.

12.3.4 *Suitability for a room's use in a recording project*

The best rooms for tracking are not always in perfectly set-up recording studios. It's a shame, but many great recording spaces exist where there is no easy possibility to rig up 16 microphone channels and perform the recording. Churches, concert halls, and school and college recital rooms can make great recording spaces and can often be booked for a cheaper price than a recording studio space too. What you'll find with a recording studio, however, is everything you need for the session to run smoothly and without compromise, with acoustics that have been tried and tested, with cable lines running directly to the mixing desk, built-in headphone foldback channels, and a set-up of professional equipment that is usually reliable and kept in good working order. To record in a non-studio space takes much more commitment to building a portable recording rig, experimenting with the set-up and room acoustics, and being prepared for unknown interruptions and issues. That said, many great recordings have been made outside of a conventional studio space, so the decision of where to record may differ depending on many factors. One great benefit to using a professional studio space is the ability to review and evaluate the recordings while they are in progress. It is extremely beneficial to be able to listen immediately back to a drum sound recording in a controlled acoustic environment, so that a critical judgement on the work in progress can be made. This is really not possible with most portable recording projects, so experience and best judgement must be relied upon instead. In particular, it can be very difficult to judge the actual sound of the kick drum recording in a non-isolated recording space. Often, even with middle-tier recording studios, some of the natural kick drum sound bleeds through the walls into the control room while you are listening to the live performance, making it sound deep and powerful. Yet, when you play back the recording, much of the deep, powerful kick drum sound is lost, because it wasn't actually captured in the microphones during the tracking; you were just fooled by the imperfect sound isolation between rooms! Such subtle benefits of using a professional recording space make a significant difference, in not only enabling the capture of great sound but also enabling the critical evaluation and essential on-the-spot decision-making that is required during a recording session. You should also never underestimate the benefits of having a non-studio space for musicians to relax and talk in without disturbing the ongoing project, and having a good coffee shop nearby too!

12.3.5   Positioning drums within a room

Once the right room and recording space have been sought, deciding where to place the drums within the room is still an important task. Even something as simple as placing a large heavy rug or drum mat on the floor can soften the reverberation coming from a hard tile or wood surface. Equally, there is usually one space in the performance room where the sound of the drums just projects better than everywhere else; usually it is in the middle and fairly central, but often a slight off-centre placement or at an angle to the walls can have a positive effect on avoiding any room mode peaks or nulls. Grammy winning producer/engineer Mike Exeter (Black Sabbath, Judas Priest) describes a valuable technique for identifying the best place to position drums within a room, by walking around with a floor tom in hand, hitting the tom, and searching for the position that provides the most impactful sound of the drum. Mike explains

> I want to find out how the room and drums react together. We can generally compensate at higher frequencies for reflections and harshness/

dullness issues, but it is a lot harder at the lower frequencies. The floor tom is the most resonant and deep sounding element in most kits so by walking round the room banging it with a stick you get to learn how the room reacts at the lower frequencies. I am looking for as much sustain and body as possible – this indicates the least amount of phase cancellation at the lower frequencies. Once I find this spot I get the drummer to set up their kit around that tom.[8]

Circular rooms can cause a particular problem, because there is a build-up of sound reflection at the exact middle of the space, so you should avoid situating the drum kit directly in the centre of a circular room. If you are in a circular room, walk to the centre, and clap your hands, you'll most likely hear a cascading *flutter echo* of the sound – it's a pretty amazing acoustics phenomenon, but not an effect you want to hear in your drum recording! It's therefore clear that experience of acoustics and listening is a valuable asset when deciding on a room and drum kit positioning for recording, and it's particularly valuable to evaluate some of these things before the day of the recording if at all possible.

TRY FOR YOURSELF: EVALUATING RECORDING ROOMS

With a low-frequency floor tom in your hand, and a stick or mallet, walk around a drum recording room or practice space hitting the drum. Listen particularly to the lowest frequencies in the drum sound and observe the positive and negative characteristics of the sound. Some positions in the room will sound thin and weak, whereas others might sound overly boomy and resonant. Other areas might sound a bit unclear and muddy, but another spot might have the perfect balance of power and clarity. This is a great way to decide where to place the drums in a room for recording.

12.4 Tuning and performance for recording

Setting up, tuning drums, and performing drums for a recording project is subtly different to tuning and playing drums for a live performance. In a live performance, we want the sound of the drums to blend well with the other instruments in the band, to be balanced in volume but also in reverberation, so that the drums do not wash and drown out the other musicians, or lose impact and directness for the audience. This is influenced significantly by the room and the mixing and amplification equipment in a performance space. If no amplification of the drums is provided, necessary or possible, then the drummer may have to tune the drums very specifically with respect to the performance space. If the room has a large reverberant character, then extra damping might need to be added to stop the drums ringing out too long. It's usually not possible to change drumheads to a more damped type at short notice, so you may be caught without the perfect drumheads for the space you are playing in and revert to some external or retrofit damping solutions instead. If the performance stage allows close microphones to be positioned on every drum, then this is less of a concern for the drummer, since generally the drums will be in a space or venue that is too big for their acoustic sound to project sufficiently to cause detrimental reverberation characteristics. In this instance, the live sound engineer has the ability to create a mix of the drums from tight close mic sound sources and add their own artificial reverb to the drum sounds if required, or leave the artificial reverb set to a lower value if the room has its own significant reverb characteristic. These live performance situations described are very different to the studio performance situation, where every effort can be made to set the drums up perfectly for the space; to choose the best drumheads; to set up microphones, drums, cymbals and other accessories to achieve exactly what is desired; and to change the set-up for each song being recorded if required. While it's advantageous to consider all this in pre-production, it is still valuable to take the necessary time during the recording session to achieve the best and most appropriate recorded drum sound for each song being tracked.

Every song in a recording project has a different feel, tempo, musical key, and possibly genre too, so there are many reasons why you might choose to modify the recording set-up for some songs. There might be some songs which require a tight high-pitch sound from a metal shell snare to cut through the mix. Others need a warm low tuned snare sound, maybe from a different snare drum. Some songs use lots of toms, whereas others don't use any – it can be advantageous sometimes to remove unused drums from the set-up if not required, so that they don't add any extra ringing or vibration to the overall kit sound. Some songs might have a musical key that clashes with the tuning of the toms, so a retune could be required. Sometimes drummers play so hard in the studio that drumheads need replacing, and it's valuable to retune them to an exact benchmark frequency and resonant tuning factor to keep consistency as the project progresses. It's also possible to identify that some songs, at slower tempos, allow space for the drums to be tuned with less damping and longer decay times that ring out quite musically, whereas other songs may be at a fast tempo and require more significantly damped drumheads to get the tight sharp sounds that are desired. So there are many considerations for the drum set-up during a recording project, and an appreciation of these allows a collection of songs to sound cohesive, yet varied and in their own context as well, when listened to as an album or playlist.

There are also many tricks and techniques that drummers can use in the studio, which just aren't possible to organise during the fast-paced adrenaline of a live performance. For example, it's possible to experiment with sounds; perhaps placing a cloth over the rack tom gives it a unique and authentic sound that perfectly suits a more retro-influenced song. If the kick drum isn't sounding sharp enough or with enough attack, a piece of plastic can be taped to the drumhead where the beater hits to give a subtle click sound to the impact. Conversely, it's also sometimes the case that a softer performance in the studio can be mixed to sound bigger and more powerful once *dynamic range compression* and *spectral equalisation (EQ)* is added – so even the performance style can be experimented with in a way that just isn't possible in a live performance. There are no creative rules in the studio: if it sounds good, then it is good – just be careful not to break anything and to not waste too much time experimenting that you only get half the songs recorded that you intended to!

Notes

1 The many roles and responsibilities of the music producer are discussed thoroughly in *The Art of Music Production: The Theory and Practice*, 4th Edition, Oxford University Press, 2013, by Richard James Burgess, who is a multi-award-winning music producer and an incredibly experienced drummer too.
2 The many components of pre-production are discussed in detail in *What Is Music Production? A Producer's Guide, the Role, the People, the Process*, Focal Press, 2011, pp. 141–58, by Russ Hepworth-Sawyer and Craig Golding).
3 As mentioned by Rob Toulson in *Sounding Off: Think One Step Ahead*, Sound on Sound, March 2010, p. 202. Available online at https://www.soundonsound.com/people/sounding-think-one-step-ahead [accessed 01/08/2020].
4 The Good Rule is devised and discussed by David Miles Huber and Robert E. Runstein in their book *Modern Recording Techniques*, 9th Edition, Routledge, 2017, p. 106.
5 Article by David Greeves detailing Tommaso Colliva and Alex Reeves' drum room investigation, *Recording Drums: What Difference Does the Room Make? One Drum Kit. Seven Rooms*. In *Sound on Sound Magazine*, September 2016. Article and example recordings available online at https://www.soundonsound.com/techniques/recording-drums-what-difference-does-room-make [accessed 01/08/2020].
6 For example *Master Handbook of Acoustics*, 6th Edition, McGrawHill, 2014 by F. Alton Everest and Ken C. Pohlman.
7 The inverse square law is described in studio recording context in *Sound and Recording: Applications and Theory*, 7th Edition, Focal Press, 2014, pp. 18–19, by Francis Rumsey and Tim McCormick.
8 Interview with music producer/engineer Mike Exeter conducted on 21/07/2020.

13 Fundamental technologies for drum recording

When recording any instrument, the equipment used to capture the recording has a big influence on the resultant sound when played back. The first recording systems were entirely *analogue* and those, such as the *multitrack reel-to-reel tape recorder*, are still considered to contribute a beneficial sonic quality to music recordings. Nowadays we are predominantly interested in *digital* recording with a *digital audio workstation* (or *DAW*), in order to benefit from features and technologies which can only be enabled on a computer-based system – such as audio storage drives, immediate recall of settings, online file sharing, and the use of advanced editing and sound processing software. But sound itself is inherently a physical or analogue quantity, so most recording-related set-ups incorporate a combination of analogue and digital electronics, in order to convert the sound pressure waves into digital numerical data that can be utilised on a computer. We therefore have a recording signal chain of tools and equipment, and each component in the signal chain has an influence on the quality and the character of the recorded sound. This is relevant particularly for drums, because drums have both a wide *dynamic range* (i.e. they can create very loud and very quiet sounds) and a very *transient* profile, which means they can change from quiet to loud very quickly. These two cases push many pieces of audio equipment to their limits of performance, so it's no surprise that audio engineers often evaluate new pieces of equipment in a drum recording session.

It's not the purpose of this book to give a complete overview of all recording techniques and technologies, and there are many excellent books that already do this very well. This book therefore gives a more specialist discussion of essential information and highlights those aspects of recording, mixing, and production which are most relevant to drums. If you are interested to learn about the wider uses of recording equipment and music recording techniques in broader detail, then you may want to look at books by Corbet,[1] Huber and Runstein,[2] and Savage[3] to cover all instruments and more related topics.

13.1 Microphones and transducers

There are a number of ways to convert sound pressure variations into electrical signals, and so there are a number of different microphone (or *mic*) types available, each with its own characteristics. Through knowledge and experience it is possible to see that different methods of converting sound energy to electrical energy (i.e. different microphone designs) are more suited to some applications than others. So if you are recording drums, it's valuable to know which types of microphones will give the best results in different scenarios. Things that are common to most microphones are the fact that they usually have a very light *diaphragm*, which is sensitive to disturbances in sound pressure. The internal electronics of the microphone measure the disturbance of the diaphragm and convert this to a corresponding electrical voltage, which can be processed further by analogue or digital audio equipment.

13.1.1 Dynamic microphones

The *dynamic microphone* works on the theory of *electromagnetic induction*, which states that whenever a conducting metal cuts across the flux of a magnetic field, an electric current that is relative to the magnitude and direction of the conductor's motion is generated within the conductor. This is essentially a loudspeaker in reverse, relying on a wire coil that moves through a magnetic field when the diaphragm experiences sound pressure variations. Dynamic microphones are strong and rugged and can handle high sound pressure levels, as well as a bit of light user abuse. The magnet and the internal moving parts, however, are quite large and heavy in comparison with other microphone types, so the dynamic mic does not have such a fast response to sound pressure variations. This makes the dynamic microphone a little inaccurate, as it sometimes cannot respond quickly enough to rapid changes in sound pressure. The recorded sound of a dynamic mic is good

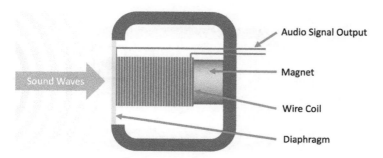

Figure 13.1 Simplified dynamic microphone schematic.

quality however, particularly when recording loud, low-frequency instruments such as drums, bass guitar, and electric guitar.

13.1.2 Condenser microphones

Condenser microphones work on the *electrostatic principle,* which defines how a capacitor stores an electrical charge. (Note that *condenser* is an old word for *capacitor* – they mean the same thing, but the word *condenser* has stuck when referring to this kind of microphone.) A capacitor is an electrical device capable of storing an electrical charge between two very thin plates. The capacitance is determined by the properties of the materials used as well as the surface area of the plates and the distance between the plates. A condenser mic therefore uses a capacitor with one fixed plate and one mobile plate; the mobile plate is a diaphragm which reacts to sound pressure disturbances. The condenser mic also needs power to supply a constant electric charge to the capacitor plates (which we call *phantom power*) and a very large resistor to measure the voltage changes across. The resistor needs a very high impedance (resistance) to avoid the circuit causing adverse filtering to the audio signal, but this results in the voltage output being very low level across the resistor. So, with a condenser microphone, we need an amplifier circuit to raise the signal level and reduce the impedance of the actual output signal. The distance between the condenser and the signal amplifier needs to be as short as possible to avoid hum and signal losses, so the amplifier is usually contained within the microphone. Amplifier circuits can be designed with *transistors* or analogue *vacuum tubes,* both requiring external power to operate (so this is taken from the phantom power supply also). Vacuum tubes (for mics and instrument amplifiers) are particularly highly regarded because they generate a pleasant form of analogue distortion often referred to subjectively as "warmth". But valves are very old and retro technology, so nowadays valve audio devices can be relatively expensive. The design allows

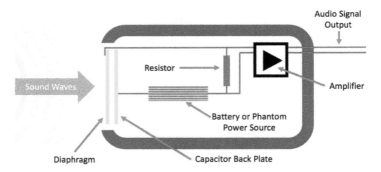

Figure 13.2 Simplified condenser microphone schematic.

condenser microphones to have a very lightweight diaphragm, and hence are extremely accurate across all frequencies; they are however quite delicate and easy to damage.

13.1.3 Ribbon microphones

Ribbon microphones are effectively dynamic microphones (as mentioned earlier, which work on the principle of electromagnetic induction). The difference is that ribbon mics use a very delicate ribbon diaphragm which responds to sound pressure disturbances. Where a standard dynamic microphone has multiple coils of wire that cut the magnetic field and generate an electric current, the ribbon uses a single, thin metallic ribbon suspended in a magnetic field, acting as both the diaphragm and the conductive element. Motion of the ribbon backwards and forwards within the magnetic field induces a small voltage that is equivalent to the sound pressure disturbances around it. The voltage is very small and often a *step-up transformer* inside the microphone is required to convert the signal to a voltage level that can be sent down a microphone cable without too much interference. As a result, ribbon microphones are very responsive and accurate, but have the downside of being relatively noisy (generating more hiss) in comparison with dynamic and condenser microphones. Because the ribbon

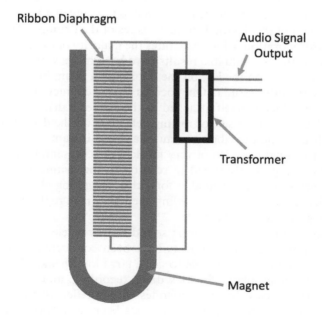

Figure 13.3 Simplified ribbon microphone schematic.

element responds to sound waves arriving from the front or back, but is insensitive to sound coming from the sides, most ribbon mics have a natural *figure-of-8 polar pattern* (which we'll discuss more in Section 13.2). The design therefore inherently makes ribbon microphones ideal for reducing unwanted noise between two instrument sources. Classic ribbon designs do not contain much internal circuitry — just the ribbon, magnets, transformer, and occasionally a passive high-pass filter circuit within. They are extremely sensitive and unsuitable for loud dynamic signals, though work well with drums as room mics and overhead microphones. Ribbon mics are generally great for giving a detailed and warm sound, owing to their lightweight and very sensitive ribbon diaphragm and an inherent high-frequency roll-off.

13.1.4 Other audio transducers

Other types of acoustic sensors or *transducers* are not specifically microphones. For example, vibration sensors using piezo-electric crystals which give off a small voltage when they vibrate. The voltage changes correspond to the type of vibration experienced, which itself corresponds to the sound transmitted through air, so we can use these (sometimes called *contact microphones*) for music recording too. The sound from a contact microphone is not hugely accurate to the sound of the instrument, because it is affected by the physical properties of the material that it is in contact with. For example, if a contact microphone is attached to the shell of a kick drum, it will not pick up all of the detailed characteristics of the drumhead's backwards and forwards vibration, which generates the majority of the sound energy of drums. Nevertheless, valuable results have been made with contact microphones, positioned on the kit, on the floor near a drum kit, or even on the walls of the recording studio. One specific type of contact microphone is the *drum trigger*, which is itself attached to the side of a drum (usually a kick or snare drum) with the sensor resting delicately on the drumhead.[4] When the drumhead is hit, the vibration causes a sharp, instant signal in the drum trigger. This signal itself is not very musical or representative of the drum sound, but it is perfectly timed in relation to the drumstick hits. Hence, drum trigger signals are valuable for either triggering electronic samples during a live performance or enabling accurate drum replacement samples to be used at the mixing stage.

There are many ways to convert sound energy to electrical energy, and many wild and wacky experiments have been conducted by studio engineers. For example, it is possible to use a large loudspeaker cone as a very low-frequency microphone for kick drum recording, and this can pick up some sub-frequencies that microphones with smaller diaphragms cannot achieve. More recently, this technique has been developed into specialist sub-frequency microphones that are commercially available and used by many studio engineers.[5]

13.2 Microphone characteristics

13.2.1 Microphone polar patterns

The *polar pattern* of a microphone defines its *directional response*. The directional response can be measured in all three axes, from the front, side, and above, as shown in Figure 13.4.

Figure 13.4 shows the characteristics of microphones with different polar patterns. An *omnidirectional* microphone has the same response for all angles; this means that wherever you point it, it will pick up sound at the same level from all angles. In practice, this is rarely 100% true, because the microphone casing itself will cause some slight isolation of sounds from directly behind or underneath the microphone. Omnidirectional mics are very accurate because they use a very simple design, meaning they work well for capturing a realistic and authentic sound of the instrument or performance being recorded.

Cardioid microphones use a cleverly designed casing to give a bias to sounds coming directly from the front – named as such because their polar pattern represents somewhat the shape of a human heart. Acoustic slots in the case's side and in the microphone capsule itself, along with greater isolation of sounds from the back of the mic case, allow a tighter directional response. Sounds from the side are picked up somewhat, but not as clearly as those from the front. Dynamic microphones are therefore good for capturing a specific sound whilst ignoring or rejecting other sounds from elsewhere in the performance space.

The *super-cardioid* design is a slightly more extreme version of the cardioid, giving a tighter sweet spot and better rejection of sounds to the side. A common trade-off is that a little more sound from the rear is captured. More extreme versions of the super-cardioid are the *hyper-cardioid* and *shotgun*

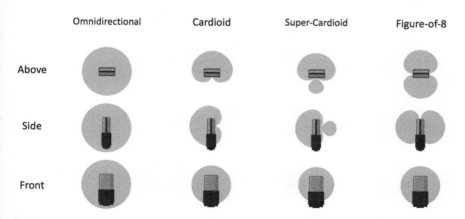

Figure 13.4 Microphone polar patterns from three angles.

microphone, which are intended to give very tight directional response fields and are used for capturing specific sounds that may be at a distance away from the microphone itself. For example, hyper-cardioid and shotgun microphones are ideal for wildlife recording and for recording an actor's dialogue in a film or TV production, but they can also be useful with drums in some circumstances where very clear individual sounds of each drum or cymbal are required.

The *figure-of-8* or *bidirectional* microphone captures sound equally from the front and rear, but has very good rejection at the sides. The ribbon in a ribbon microphone is suspended to only vibrate backwards and forwards when it picks up sound, so naturally it works with a figure-of-8 profile.

More modern microphones actually have two moving microphone diaphragms within – one facing backwards and one facing forwards, and with a shared backplate in between.[6] This creates two cardioid responses and allows different levels of signal blending between the two diaphragm outputs, to create a number of different polar patterns. So a single microphone can be designed to have selectable patterns, which can be switched depending on the particular recording application. The theory of blending polar patterns with the electric signals from two cardioid mic capsules is shown in Figure 13.5.

To complicate things, with most microphones, the polar pattern changes over the frequency range, and a microphone may have a perfect cardioid response at 500 Hz, but this might change slightly to a super-cardioid pattern at 5,000 Hz. For this reason, microphones often record a slightly different sound characteristic depending on the angle at which they are positioned to the source instrument.

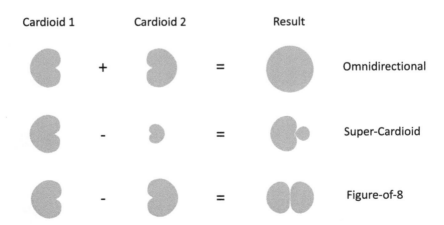

Figure 13.5 Creating polar patterns by electronically blending two opposite faced cardioid signals.

> **TRY FOR YOURSELF: EXPERIENCE DIFFERENT MICS AND POLAR PATTERNS**
>
> It's extremely valuable to experience the difference between different microphone types, polar patterns, and microphones from different manufacturers too. If you have access to a number of different microphones, then set up a number of different types to all record the same sound source, with some pointing at a different angle to the sound source too. You could use a snare, floor tom, or a hand percussion instrument such as congas or bongos. You could even try with an acoustic guitar, guitar cabinet, piano, or any other instrument. The purpose of the exercise is to understand how different microphones act and how much influence the microphone type and polar pattern has on the recorded sound. Critically listen back to the recordings and identify the qualities and weaknesses of each microphone recording.

13.2.2 Other microphone characteristics

There are a number of other characteristics relating to microphones which are worth understanding and experiencing the sound of. These characteristics include diaphragm size, *frequency response*, and *transient response*.

Diaphragm size

Larger diaphragms capture more sound energy, and hence generate larger voltages, in a similar way to how larger drums emit more sound power. The larger voltage means that the signal is generally cleaner and less affected by noise or hiss than microphones with smaller diaphragms. This is important when recording very quiet sounds to enable what is called a good *signal-to-noise ratio*. The advantage of smaller diaphragm microphones is that they are less susceptible to internal acoustic frequencies. All microphones have an internal *resonance* or *natural frequency*, in the same way all drumheads have a fundamental frequency. With small diaphragm microphones, this resonance is very high, higher than humans can hear. But with large diaphragm microphones, the resonance is often within the human range of hearing, so larger microphones tend to add their own unique character to the recorded sound.

Frequency response

We mentioned above that larger diaphragm microphones have a frequency characteristic related to the natural frequency of the diaphragm. This might be great if you are looking for that exact characteristic, but if you are looking for a very accurate recording of the sound, then smaller diaphragm microphones may be more suitable. We judge a microphone's frequency response by looking at how equally it captures different frequencies, and most manufacturers publish this data on their websites. For drums it's important that microphones pick up low frequencies very well and extend down to the lowest frequencies we can hear, which is not so important for recording violin for example. In general, we want the frequency response of a good microphone to be as flat as possible and extend over the full range of human hearing, though it's acceptable that some large diaphragm microphones have a small peak somewhere in the range, which relates to the natural frequency of the microphone itself.

Transient response

Transient response is a measure of how quickly a microphone's diaphragm can respond to changes in the acoustic sound pressure waveform. This can lead to wide differences in the resulting sound between microphones, particularly when used with percussive sounds. It's no surprise that generally smaller diaphragms, which are less heavy, can sometimes respond quicker than larger diaphragms. We obviously want a microphone to respond very quickly to drum sounds, which are one of the most dynamic instruments that generate rapid changes in air pressure. It's hard to measure the transient response of a microphone in a quantity that is meaningful to the creative process of sound recording, but through experience it's possible to identify the most responsive and accurate microphones for recording drums.

13.3 The complete recording signal chain

The microphone is just the first electronic device in the recording signal chain. There are many ways to record sound into a digital audio workstation and many available resources on this extensive topic. Figure 13.6 shows an example DAW recording signal chain that is commonly used for drums. In general, there will be many microphones and one *pre-amplifier* (or *preamp*) per microphone. The preamps may be part of the *mixing desk* which allows signal manipulation, monitoring, and routing to the *audio convertor*, which allows audio signals to be processed by a computer DAW system. In modern portable systems, the functionality of the preamps, mixing desk, and audio convertors may even be combined into a single *audio interface* device, combined with a software-based configuration and control program.

Figure 13.6 Common microphone recording signal chain.

Modern microphones use *balanced* signal lines for carrying the audio, incorporating a ground, positive, and negative form of the audio signal. This allows noises that might be induced in the signal cable to be cancelled and minimised through some well-designed electronic circuitry at each end. As a result, most audio signal connections use balanced (three-pin) connections and are commonly referred to as *XLR* (meaning *external line return*) cables.

13.3.1 *Audio convertors*

The audio convertor or audio interface is not the first component in the recording signal chain after the microphone, but it is an essential component for any computer based recording. The audio convertor's role is to convert the analogue electronic signal into equivalent digital values that a computer can use for sound processing. This is a very non-trivial task and different *analogue-to-digital convertors* (or *ADCs*) perform very differently. For example, timing is critical, since every audio sample need to be converted very accurately and rapidly at perfectly equal intervals, which pushes the limits of digital technologies. There are different calculation methods for converting analogue signals to digital values too, different qualities of electronic components, and many different approaches for ensuring that unwanted sound artefacts are not created as part of the process; for these reasons, there can be a huge range in price and quality between audio interfaces. Indeed, when testing the quality of audio convertors, drums are usually the instrument of choice because they push the limits of dynamic range and rapid transient detail that the system must handle, so it's no surprise that the convertor used can have an influence on the sonic quality of recorded drum sounds. Nevertheless, technology has evolved in recent years, and many budget level audio interfaces perform very well for their cost, meaning that anyone can engage in drum recording and can expect to achieve a professional-level sound if care and attention is taken in the recording process.

The audio convertor also has the role of converting digital sound back into analogue so that we can listen through loudspeakers or headphones, or perform other forms of analogue audio processing. This conversion (*digital-to-analogue conversion* by a *digital-to-analogue convertor*, or *DAC*) is also equally as challenging in electronic design and is of huge importance since we will listen to and evaluate our recordings based on this conversion process.

13.3.2 *Mixing desk*

When tracking multiple microphones all at once, as is the case with drum tracking, it is extremely advantageous to use a mixing desk to control and operate the studio session. The mixing desk is designed to make as many options as possible available to the recording engineer. Nowadays, if recording to a DAW, the mixing desk can be less critical, because we are able to leave some decisions about volume and balance specifically to the mixing stage. The DAW allows many audio tracks to be recorded in with very few limitations, so the mixing desk's role is to allow the session to run smoothly and efficiently, and to ensure that all signals captured into the computer through an audio interface are clean and accurate. The large-scale mixing desk looks daunting to an inexperienced user, but predominantly the desk includes lots of repeated functionality for multiple audio channels, so it is possible to break down the features to more understandable aspects. As a drummer, it is certainly worth understanding the basic features of a mixing desk and understanding its role in the recording process. This is best done through practical experience and perhaps shadowing a recording engineer while they work, if the opportunity presents itself.

13.3.3 *Microphone preamplifiers*

Most mixing desks have microphone preamplifiers installed, which are essential for recording microphone signals, though bespoke and special-ist preamps can be used too, either in combination with a mixing desk or directly into an audio interface. Many audio interfaces have preamplifiers built in too, allowing a mixing desk to be omitted on less complex recording projects.

All microphones give out very weak voltage signals. The first thing that's needed for the microphone signal is a little amplification, so that it can be used in a mixing desk or passed through an audio convertor without be-coming noisy or lost. A microphone preamp is therefore a low-power audio amplifier that takes a microphone signal from a level of millivolts to usually somewhere in the 0–5 V range.

The second role of the preamplifier is to provide phantom power to any condenser microphones that are being used. The quality of the phantom power source is critical to getting the best performance out of a microphone. A preamplifier uses an electronic *rectifier* circuit to convert *alternating mains current (AC)* into a flat *direct current (DC)* that can be used to power a condenser microphone. This rectifier conversion is therefore critical to the quality of the signal being recorded and the stability of the amplified signal that is forwarded to the mixing desk or interface. All preamplifiers therefore add their own sonic effect to the signal, since they have different internal noise tolerances and they have their own unique frequency responses. Most professional mixing desks have preamps on the input of each channel; how-ever, many producers choose to use specialist *outboard* preamps to provide the amplification and leave the ones in the desk set at fairly low gain, or use a desk with high-quality preamps installed.

You might ask "what's all the fuss about with preamps?" Well, you can only really decide if expensive preamps are worth the money by listening. However, there is a whole body of evidence that the use of good preamps can be the difference between a good sound and an excellent sound, and it's worth reading the recommendations from top producers and recording engineers to understand their personal preferences and the importance they put on preamps, and the analogue electronics that come before the audio signal is digitised.[7] Drummer and music producer Emre Ramazanoglu emphasises the importance of good-quality microphone preamps and shares his own recording signal chain that includes some specialist analogue tape emulator circuits:

> The preamps for recording drums are very important. The API preamps react very fast and allow all the detail of the drum sound to be recorded into the DAW. I also use the Roger Mayer 456 Tape Emulator units on each channel after the preamp. They give me the one thing about analogue tape sound that works great with drums. They control the peaks and give what I call "peak efficiency" without changing the effect of the transient, so they allow the blend between the attack and the tone of the drum to be more balanced.[8]

13.3.4 Recording with signal processing effects

The mixing desk also enables the opportunity to implement signal processing effects at the point of recording. Whilst this might not be necessary and can be left until the mixing stage, you may want to add some equalisation, compression, or other audio processing to the drums as you record them, to create a more finished sound and to give a better understanding of the sonic character which is being captured. Similarly, you might add some reverberation at the recording stage purely for the drummer or other band members who like to hear a more finished style of sound in their headphones while tracking. The mixing desk allows this by enabling grouped or *auxiliary* sends of each microphone channel to external audio equipment, which can then be brought back into the desk and processed on a new mixing channel specifically for that effect. To avoid duplication of information in this book, we will only discuss audio processing for drums particularly as a post-recording process in Chapters 16 and 17; however, many of those effects discussed can be implemented at the recording stage if desired.

13.4 Monitoring and foldback

13.4.1 Control room monitoring

When recording drums, or any instrument, it's important to be able to evaluate the sound that is being recorded at the time of the recording. The drums may sound fantastic when you are stood next to them in the performance room, but the final recordings may have much less appeal, sounding muddy, boomy, overly reverberant, or just plain boring. It is essential therefore to be able to evaluate the recordings before the end of the recording session, and ideally before you do the first proper take of a song. When playing back a test recording, the types of loudspeakers you use to listen back or *monitor* the recordings have a significant influence on the sound too. Look back to one of the very first diagrams we presented in the book (Figure 2.1) to remind yourself how the speaker, the transmission space, and our hearing can all affect the quality and our judgement of the sound being played back. For this reason, engineers prefer to use a control room space with professionally designed acoustics and high-quality *reference monitors* (loudspeakers), so that they can quickly decide if the drum set-up and microphones are capturing the desired drum sound. Don't underestimate the value of taking some time to evaluate the sound as you record. If there is an issue, then a whole day or more of recorded drum tracks may be totally unusable when you evaluate them at the mixing stage. Experience with different equipment and recording spaces can make this quite a straightforward task after a few sessions, but all engineers most likely have a story of how they spent a day recording only to realise afterwards that the material was not suitable for mixing, owing to the wrong drum tuning for the song style, a constant buzz or external noise on the recordings, distortion that wasn't obvious during tracking, too much room reverberation, too much bleed between microphones, or maybe a microphone that was broken or just didn't record properly.

We've already mentioned how important it is to have total isolation between the control room and the performance room too; but sometimes this just isn't possible. Maybe the recording studio is not perfectly built for isolation, or maybe you are recording in an ad hoc space and sat in the same room as the drummer and drum kit. This is not ideal, sonically, but can be managed, especially with the use of very good-quality headphones for reference. (It must be noted that there are sometimes more practical benefits of being in the same room as the musician, and some engineers and producers embrace this type of set-up.) Whilst it may not be possible to fully evaluate the drum kit's sound as it is recorded, with good-quality headphones it is possible to listen back to some test recordings and make a judgement on how well the drums are set up. In this case, it's valuable to own a set of professional headphones that you have experience of listening to many different types of music on. If you know critically what your favourite songs sound like in your own pair of studio headphones, then you are in a good position to make quick judgements on drum sounds as you record them.

13.4.2 Headphone foldback

The *headphone foldback* or *cue mix* which a drummer listens to during a recording session is very impactful on their ability to perform well. As music producer/engineer Mike Exeter explains:

> If you make the musician feel special they will perform as if they are playing to the biggest, and most supportive, audience in the world. When

you work with them to present them with a fantastic sound through their cue mix, they know you are invested in making them sound incredible and giving them the best recording experience they could want.[9]

Unfortunately, setting up a headphone mix for a musician is rarely a simple task and requires some specific time, care, and attention. However, when done well, this can transform a challenging recording session into one which runs smoothly and is a joy for all involved. This is specifically relevant when recording drums, since the drummer usually needs to hear a clear metronome click, whilst also listening to the performances of other musicians in the room, or to some previous recordings as a guide track. In either case, if the drummer can only hear their own drums, they will find it very difficult to lock in to the timing and feel of the song being recorded. There are many different methods for setting up a headphone mix, and the correct one for any situation depends on the available equipment and tools in the recordings session. Large format recording desks are well equipped to make a separate mix of the music to be perfectly balanced for the drummer to hear everything they need. Some more advanced headphone systems even allow the drummer to control their own mix of the backing track, the click, and their live drum sound in order to find the perfect balance from their drum seat. With more portable systems, there are often less options to create a bespoke mix for the drummer, though a solution to this should ideally be sought and time spent tailoring the foldback to the performer's preference. Many audio interfaces and DAW packages have advanced routing options, so if you have a good handle on the software and systems being used for the recording, then it is usually possible to find a good solution.

In all cases, there is a big advantage for the engineer being able to hear exactly what the drummer is hearing, in order to understand why they might be experiencing difficulties with the foldback mix. If you are able to set up a set of headphones in the control room that runs from the same channel as the drummer's foldback, then you can immediately hear if there is any distortion on the headphone mix, or if the click is getting lost under the sound of the drums or backing track. It is then possible to quickly make adjustments that will get closer to a mix that the drummer will be able to perform well with and allow them to advise on any fine adjustments they might prefer.

TRY FOR YOURSELF: PERFECT HEADPHONE MONITORING

It's important to be able to give the drummer exactly the headphone mix that they require. They will always perform better if they are happy with the headphone mix. If you are setting up a recording

(Continued)

session, identify the different methods or routing which you might use to provide a headphone mix to the drummer, then evaluate which approach allows the most control and tailoring of the sound to a drummer's potential requests. Can you do this in the recording hardware itself, or do you need to set up a software headphone channel and route that back to the drummer?

It's also good practice to be able to add some effects to the headphone mix to make the sound they hear more realistic. Can you add some reverb to the headphone mix with your set-up? This is often requested by musicians and particularly singers who want to hear their own voice folded back with some reverb effect added.

Can you add a second set of headphones, so you can hear exactly what the drummer is hearing too and quickly understand their comments if they are not hearing a mix as they would like?

Notes

1 *Mic It! Microphones, Microphone Techniques, and Their Impact on the Final Mix*, 2nd Edition, Routledge, 2021, by Ian Corbett.
2 *Modern Recording Techniques*, 9th Edition, Routledge, 2017, by David Miles Huber and Robert E. Runstein.
3 *The Art of Digital Audio Recording*, Oxford University Press, 2011, by Steve Savage.
4 For example, the Roland RT-30HR and RT-30K triggers designed for use with snare and kick drums. Detailed online at https://www.roland.com/uk/products/rt-30hr/ [accessed 01/08/2020].
5 For example, the Avantone Kick sub-frequency microphone. Detailed online at http://www.avantonepro.com/kick.php [accessed 01/08/2020].
6 For example, the AKG C414 XLS. Detailed online at https://www.akg.com/Microphones/Condenser%20Microphones/C414XLS.html [accessed 01/08/2020].
7 For example, Al Schmidt (with credits including Steely Dan and Neil Young), Phill Ramone (Billy Joel, Paul Simon, and Frank Sinatra), and Chuck Ainlay (Mark Knopfler and The Chicks) all mention the importance of the microphone preamp in Howard Massey's book *Behind the Glass*, 2002, Volume 1, Backbeat Books.
8 Interview with drummer and music producer Emre Ramazanoglu conducted on 14/10/2020.
9 Interview with music producer/engineer Mike Exeter conducted on 21/07/2020.

14 We need to talk about phase!

In the following chapters, we will discuss many more aspects of both recording and mixing drums for different genre projects. But first, there is one essential foundation concept worth discussing that crops up again and again when considering recording and mixing drums: *phase*. The concept of phase is a fairly challenging one to understand, often because it is a term that is regularly misused and misrepresented. But it applies very significantly to drums and low-frequency sounds in studio production, so it is worth discussing the very basic principles of phase before moving on to those topics. Despite being a complex subject, the concepts of phase are extremely relevant for drum recording, as Grammy winning producer/engineer Sylvia Massy emphasizes:

> The importance of checking the phase relationship between each mic cannot be stressed enough. When drum mics are out of phase, even the most powerful drums will sound papery and thin.[1]

Some of the topics mentioned in this chapter also relate to the drum recording technologies and techniques discussed in Chapters 13 and 15, so you might like to cross-reference these chapters a little if some of the discussion topics are not immediately obvious.

14.1 What exactly is phase?

The concepts of *phase alignment*, *phase cancellation*, *phase coherency*, *checking for phase*, and *phase issues* (or whatever phase-related term you prefer to use!) are particularly relevant to drum recording and production. However, the topic of phase is often very much misrepresented, misunderstood, and misinformed, so don't be alarmed if you find some discussions or articles around phase to be confusing. The term *phase* really isn't the ideal choice of word for what we are usually discussing here, but it's a clever catch-all term that people can use in the studio to describe some specific aspects related to sound. Let's begin by unpicking the term and discussing what the word *phase* really means in the context of music production and specifically how it relates to drum sound.

Here's the scientific bit; the term *phase* is a purely mathematical quantity, related to pure sine waves which are defined by the equation $y = sin(x)$. The value of x is what we call the *angle* of the sine wave, and if you type a value of x into a calculator and press the *sin* or *sine* button, it will give you the corresponding value (or result) y. If you put lots of different values of x into a calculator and note all the different values of y that are given back, and draw a graph with the x data values along the bottom and the results for y in the vertical axis, you will see a perfect sine wave, as in Figure 14.1.

Sine waves were first explored and understood by Greek mathematicians, and are related to circles and trigonometry, which is why the sine wave repeats at exactly the value 360, which is how many degrees there are in a circle. So, in multiples of 360, the sine wave repeats at 360, 720, 1,080, and so on.

The sine waveform in Figure 14.1 begins at exactly 0 degrees, but a *phase offset* or *phase angle* is present when we take the pure sine wave and shift or offset it by a number of degrees, so that it doesn't start perfectly at zero. Some examples of phase offset are shown in Figure 14.2. In relation to the pure sine wave in Figure 14.1, all of these sine waves lag behind by 45, 90, or 180 degrees, and as a result, the pure sine wave shape that we expect to see

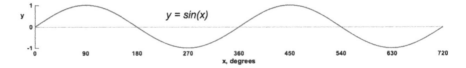

Figure 14.1 Graph of a simple sine wave with degrees along the *x*-axis.

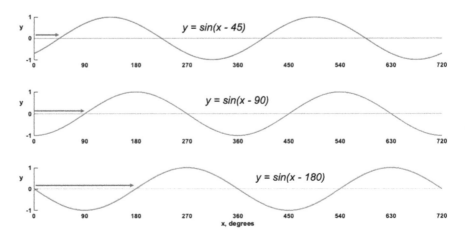

Figure 14.2 Phase offset examples.

doesn't start until the angle of the phase offset. It's valuable to note that at a phase offset value of 180, the pure and offset sine waves look like a mirror image or inversion of each other. At a phase offset value of 360, the two sine waves would look identical again, because the sine wave repeats after 360 degrees.

The concept of phase relates to audio because we can, theoretically, consider all sounds and signals to be made up of many different sine waves all added together. Equally, when we start adding audio signals together – which we do in a mixing console or a digital audio workstation (DAW), and also by particle collisions in a real acoustic space – then it is possible for similar signals to either add together or cancel each other out if they have a notable phase difference. Hence we can alter, enhance, or even lose qualities of the sound we are trying to record or play back if phase offsets are present. Some examples of mixing phase offset sine waves are shown in Figures 14.3. We can see that mixing (also *summing* or adding) two identical signals that have no offset results in amplification (the result of the summation is double the amplitude of individual input signals). When mixing two identical sine wave signals with a 180-degree phase offset, we get total signal cancellation (i.e. no sound or activity from the resulting signal). When the phase offset is in between those two extremes, we get either some amount of amplification or an amount of attenuation and a phase offset in the resultant mixed signal.

Because of the association between sine waves, music, and audio signals, the term *phase* has been adopted as part of the language of recording and music production. But, actually, there are some much more useful terms related to phase which are more appropriate to use and which make far more sense to any musician or sound engineer; and those terms specifically are

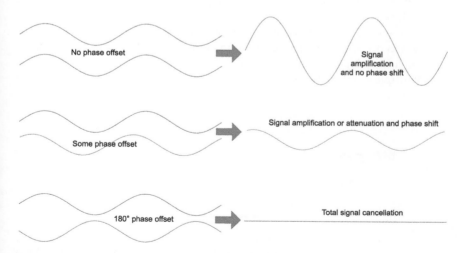

Figure 14.3 Summation of phase offset sine waves.

time delay and *signal polarity*. Usually when people talk about phase in the recording studio, they are actually referring to something associated with either the polarity of a signal or a time delay.

14.2 Time delay and comb filtering

We have seen that mixing two identical signals that have a phase offset can, to some degree, amplify or cancel each other out (this is known commonly as *constructive* and *destructive interference*), so it is useful to appreciate how this occurs in the context of the recording studio. There are a number of ways in which we experience this phase offset when recording, and the most common is owing to time delays in the recording and production process, symbolised by Figure 14.4.

There are many occasions where we might accidentally delay a sound signal and then later mix it together with another version of itself (that is either not delayed or delayed by a different amount). Hence, there is opportunity for signals to be amplified or cancelled unnaturally. In reality, time delays are impossible to avoid, since they occur in the following scenarios:

- When the same sound source is recorded by two different microphones that are at different distances from the source
- When the same sound source is recorded by two microphones that have different internal sound response and conversion characteristics
- When a sound wave is reflected off a surface in the studio and hence captured twice (or more as an echo or reverberation) into a single microphone
- When a recorded signal is split in a mixing desk (e.g. sent to an auxiliary channel for adding effects or other processing) and then mixed back with the original signal
- When a recorded signal is split in a DAW software package and processed differently (e.g. one version is sent to an effects plug-in, whereas the original is not), and then later mixed back together
- Multiple combinations and permutations of the above scenarios acting together

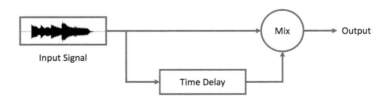

Figure 14.4 Time delay and summation (mixing) of identical signals, caused by acoustic, electronic, or computer processes, resulting in phase incoherence and comb filtering in the output.

Because the above scenarios are all common to nearly every recording project, eradicating the effect of time delays in audio is impossible. But having an awareness of specific scenarios – those which can have the most negative effect on a recording – can help us to avoid, check for, and protect against issues with phase.

Firstly, and perhaps most significantly, is to avoid frequency cancellation owing to time delay (often referred to as *phase cancellation*) caused by recording the same sound into two microphones at different distances from the sound source. This is extremely common in drum recording, particularly where there are multiple microphones set up around the kit to capture different qualities of the drum sound, and this can impact the quality of the recorded sound when all microphone inputs are mixed together. It is common to use a number of microphones at different positions, to enable greater clarity or control of each individual drum, but we cannot acoustically separate any of the sound sources, so every drum is captured in every microphone to some degree, causing multiple opportunities for frequency cancellations. Time delays are most apparent when mixing close microphone sounds with overhead or room microphones that are positioned much further away. For example, we might have a microphone 5 cm away from the snare drumhead, but that same snare sound is also captured very clearly in the overhead microphones which might be 1 or 2 m away.

Figure 14.5 shows a drum kit with three microphones. When considering the snare drum, we can see that the sound travels different distances to each microphone and hence takes different amounts of time. A DAW recording of a snare hit with this set-up is also shown in Figure 14.5, with points of interest highlighted. The snare sound is recorded in the waveform timeline at point (a), whereas the overhead mic and the room mic are

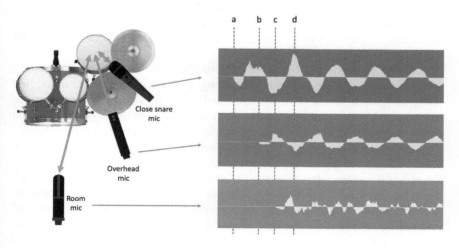

Figure 14.5 Microphones at different distances from the snare drum on a drum kit.

delayed by a small amount and start at points (b) and (c) on the timeline, respectively. A point of interest is at timeline point (d), where the close snare mic shows a signal peak, yet the overhead mic shows a signal trough. If these two signals are mixed together, then we can expect significant attenuation of the snare drum's fundamental frequency, resulting in a weak and thin sound, even if the microphone sounds are good when listened to individually.

There's a very useful number to introduce here, and that is the speed of sound, which determines the time difference between the two recordings, based on the distance difference between the two microphones. The speed of sound is around 340 m/s (it differs very slightly at different altitudes and different temperatures), which means that sound travels 1 m in about 3 ms. So if an overhead or room microphone is 1 m away from the snare drum, then the recorded sound of the snare will have a 3-ms time difference between the close microphone and overhead or room microphone data. This could cause significant frequency cancellation of aspects of the snare sound (it could actually cause frequency amplification too and maybe even make the snare sound better, if you're having a lucky day!). The reason we use the term *frequency cancellation* here is because a single time delay in milliseconds results in very different phase cancellations for each frequency in the audible range. A bit of maths shows that a 3-ms time delay causes cancellation of 167-Hz frequencies,[2] which is around the kind of frequency that some snare drums have their fundamental pitch tuned to. The key with recording is to understand which frequencies are most important to the sound of the instrument and then listen carefully before recording to make sure that these particular frequencies are not negatively impacted by the recording process. Particularly with recording drums, the most important frequencies to check for issues or phase coherency are those associated with the fundamental frequencies of each drum in the drum kit. Thankfully, having read the previous chapters, you know all about the fundamental frequencies of drums and can use this knowledge now to your advantage when recording!

The best method to get good phase coherency results, without spending many hours on setting up before recording, is to check each close drum mic separately when mixed with the overhead mics. While in the control room listening to the microphone inputs, start with the bass drum – solo the bass drum microphone and also solo one or both of the overhead microphones. Now press the mute button on the overhead mic channels and think about how the sound differs when the overheads are added back. Does anything change in the sound that you don't like? Specifically, we know that the kick drum has a main fundamental frequency at around 60–80 Hz, so listen to how this lower frequency range improves or diminishes when the overhead microphone is added to the mix. If you're not sure, you can add a low-pass filter to each microphone input to help decide, put the low-pass threshold down as low as possible, to around 500 Hz maybe, and filter away all the high-frequency detail – now you are just listening to the most fundamental frequencies of the drum and its vibrating drumheads. If there is frequency cancellation occurring, you will certainly be able to hear it now, as when you add the overhead mics to the mix, it is very common for the most powerful aspect of the drum sound to disappear or diminish. If this is the case, the solution is to move the kick drum microphone, maybe just a few centimetres will be sufficient for the frequencies to fall back into line and result in that powerful fundamental coming through strong and clear when all microphones are up in the mix. If the problem seems to persist, maybe you should move the overhead microphones a little also or instead, and see what difference that makes too.

It's wise to perform the same process on all of the drums when mixed with the overheads. It's also important to perform this if you are using more than one microphone positioned on a single drum. For example, you might have two close microphones on the kick drum to capture different sonic qualities of the sound. Be careful, because even if the two microphones are at exactly the same distance from the drumhead, this does not mean there won't be any frequency cancellation – different microphones respond in very different ways and with different internal characteristics, so you will need to check for time delay or phase coherency regardless.

TRY FOR YOURSELF: EVALUATING PHASE COHERENCY ON A DRUM KIT SET-UP FOR RECORDING

In your next recording session, follow the procedure above to check phase coherency between the following combinations:

- Overheads and kick drum: check for phase coherency.
- Overheads and snare: check for phase coherency.
- Multiple microphones: ensure phase coherency for all drums with multiple close microphones.
- Stereo overheads: ensure phase coherency when mixing both overheads to the centre.

The duration of the delay determines which frequencies are cancelled and which aren't. For example, a 2-ms delay means that a sine wave of period 4 ms will cause cancellation (because 2 ms is 180 degrees of a 4-ms wave). A 4-ms period sine wave has a frequency of 250 Hz, because 250 waves with a period of 4 ms can occur in 1 s (i.e. 1,000 ms). But a 2-ms delay causes more than one frequency to be cancelled, because of the repetitive nature of a sine wave, so we see a similar cancellation effect for the 2-ms delay at 750, 1,250, 1,750, and so on. The effect of a 2-ms delay also doubles the frequency content at perfectly in-phase frequencies such as 500, 1,000,

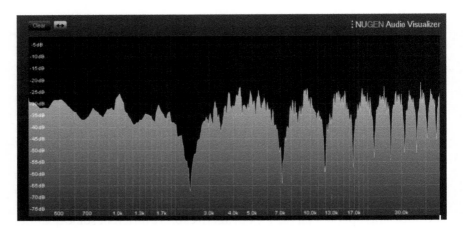

Figure 14.6 Comb filtering shown with the Nugen Audio Visualizer plug-in.[3]

and 1,500 Hz. This can also be evaluated on the same signal for a different time delay. For example, for a 6-ms delay, the frequencies of cancellation are 83, 249, 415, 588 Hz, and so on. This effect of phase incoherence is known as *comb filtering* because multiple frequencies are filtered out and the subsequent frequency spectrum resembles the shape of a hair comb, as shown in Figure 14.6.

TRY FOR YOURSELF: EXPERIENCE THE SOUND OF COMB FILTERING

In Logic Pro, set up an audio channel with the Test Oscillator plug-in inserted and set to a White Noise source, as in Figure 14.7. Now send the signal to an auxiliary bus with a 0-dB level. You now have two versions of the same white noise signal playing at the same time.

Add a sample delay to the auxiliary channel and add a one sample delay to the auxiliary channel. Can you hear the difference in the sound of the noise signal? Now increase the number of samples delayed to 2, 3, 4, 5, and so on, and each time notice the difference in sound when the sample delay changes. The sound you hear is the difference in sound caused by comb filtering.

If you have access to a good spectrum analyser plug-in, open this up on the master channel and you will be able to see the comb filtering in action too.

Now switch off the test oscillator and add an audio file to the source channel. You can now implement the sample delay on the auxiliary again and hear how the sound of a recording is effected by comb filtering.

Figure 14.7 Logic Pro set-up for evaluating comb filtering.

It's interesting to note that *phaser* and *flanger* audio effects use the principle of comb filtering to produce a novel and creative sound effect. Phaser and flanger effects electronically create comb filtering and then slowly change the time delay to move the cancellation up and down through the frequency spectrum, giving a "sweeping" or "phasing" sound to the audio. This explains why the sounds of acoustic comb filtering and phase cancellation differ to the phaser processing effect, because in unwanted cancellation, the time delay stays constant, so the sweeping sound is not present.

14.3 Mono compatibility

When recording an instrument in stereo with just two microphones, we can avoid phase cancellation by panning one signal fully to the left and one fully right. The signals are never mixed electronically or in software, so cancellation cannot easily occur. If the speakers are situated apart, then the chance of acoustic mixing and cancellation is limited also. However, many mobile phones, wireless speakers, cheap TVs, and portable radios have only a mono speaker, so the left and right channels of any music are mixed together before being output in a mono form. In this case, if the left and right channels are out of phase, then cancellation will occur, which usually results in a loss of bass frequencies (i.e. drum frequencies!) and, hence, a weaker, unbalanced sound that lacks power. The worst-case scenario is when the two channels carry identical signals with opposite polarities – summing the two channels to derive a mono signal would result in silence!

When setting up drums, it's valuable therefore to also check the phase relationship between any stereo microphone set-ups you have. For example, if you have a pair of overhead microphones capturing the overall sound of the kit on both the left- and right-hand sides, then it's essential to listen to the sound of those microphones when mixed together, or as we often say *in mono*. This is important because what might sound great in the control room when recording might not sound very good when played back on different speakers and particularly on mono music systems. To check for phase coherency between stereo overheads, pan both microphones to the centre (or hit the mono switch on the mixing desk if there is one), and again start by listening to the kick drum. Does the sound change significantly when one microphone is muted? If so, then the chances are one of the stereo microphones is further away from the kick drum, then the other, so try adjusting the positioning until the sound does not diminish when both microphones are heard together. You should check this for the snare drum too and again check for coherency when the two overheads are mixed together. This might cause some conflict, because having just moved the mics to get a perfect positioning for the kick drum, you now need to move them again to get perfect coherency for the snare drum – and now you need to check the kick drum again! This is why some people use a tape measure to ensure that the overhead mics are as equally spaced from the kick and snare drum as possible. It's difficult to get a perfect scenario for both drums, but it's worth evaluating the situation and making a personal judgement on what sound you are most happy with.

The phase and mono compatibility of a signal can be analysed with a *phase meter* or a *goniometer*. Phase meters are common in many audio software packages; different displays of the Logic Pro goniometer are shown in Figure 14.8. The phase meter was developed to provide an indication of the relative phase of the two stereo channels and thereby provide some way of assessing mono compatibility.

In general, the goniometer or phase meter indicates the following:

- A vertical line indicates a perfect mono signal – i.e. the left and right channels are exactly the same (Figure 14.8a).
- A horizontal line indicates that the left channel is the same as the right, but 180 degrees out of phase (Figure 14.8b).
- A perfect circle indicates a sine wave on one channel and the same sine wave shifted by 90 degrees in relation to the other (Figure 14.8c).

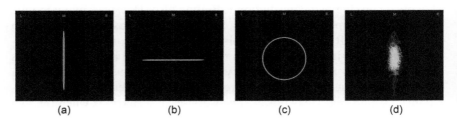

(a)   (b)   (c)   (d)

Figure 14.8 Phase meter (goniometer) showing different phase coherency readings.

- A random but rounded shape indicates a well-balanced signal. Often it is preferred for this to be around the vertical axis rather than the horizontal one, because the vertical axis represents the line of mono compatibility (Figure 14.8d).

TRY FOR YOURSELF: ANALYSING PHASE WITH THE GONIOMETER

In Logic Pro, set up an audio channel with the Test Oscillator inserted and set to a sine wave source at 250 Hz and panned fully to the left (as in Figure 14.9). Now send the sine wave signal to an auxiliary bus with a 0-dB level and pan the auxiliary signal fully to the right. You now have two versions of the same sine wave panned to opposite sides of the stereo field. Add the Logic Gain plug-in followed by the Logic Multimeter plug-in to the stereo output channel. Look at the goniometer in the Logic Multimeter, you should see a vertical line because the left and right signals are perfectly in phase. If the signal shown in the goniometer is small, you can boost it with the Gain plug-in to enlarge the visualisation.

Place the Sample Delay plug-in on the auxiliary channel and add a few milliseconds of sample delay. You should see that the goniometer becomes a horizontal line at 2-ms sample delay, indicating total mono cancellation or incompatibility. At 1 and 3 ms, you should see a perfect circle. What other time delays cause the 250-Hz signal to cancel with a horizontal line?

Try changing the sine wave frequency and the sample delay further. Can you predict what frequencies will cancel out with what time delays?

Load in a stereo audio file of some drums or a full mix and look at the goniometer readings for the audio. What can you learn from the meter displays for the real audio signals?

Figure 14.9 Set-up for phase meter analysis in Logic Pro.

14.4 Signal polarity

The first time many people in the world of music encounter phase is when they first see a *phase invert* switch on a studio mixing desk or microphone preamp, and enquire of its purpose. So what does that little switch on the desk do? You'd think that being called a phase invert button, it would invert the phase of the signal; well, actually that's a strange term in itself, because it's not actually possible to "invert" the phase of a signal! Phase can be shifted or offset as shown in Figure 14.2, and we have seen that when the phase of a sine wave is shifted or offset by 180 degrees, the result is that the signal or sine wave becomes inverted in comparison with its zero phase equivalent. But when we invert an entire signal, it means that we take all of the negative values and make them the same value, except positive. We also take all of the positive values and make them the same value, but negative. It's a strange concept but one that's employed in many audio amplifier circuits that use valve or transistor components. Using this technique results in an increase of an audio signal (i.e. amplified) but also inverts the signal within the process. A simple schematic of how the valve or transistor operates within this scenario is shown in Figure 14.10.

The solution is to add a second-stage amplifier so that the signal out of the amplifier is now bigger (amplified) but also still the same shape and profile of the original signal. We have a term to say if the signal is inverted or not, and that is described by the signal's *polarity*. If we *change, swap, flip,* or *reverse* a signal's polarity, then it becomes *inverted*.

We can see that switching the polarity of a simple sine wave is similar to shifting its phase by 180 degrees as shown previously in Figure 14.2. There are lots of ways to flip or invert a signal's polarity, we've already seen one above which is to put the signal through a simple one-stage transistor amplifier. Another is to swap the signal lines of a *balanced* microphone signal, in the connector, the cable, or the unit which the cable plugs in to. This branches into another topic on balanced microphone cables, since microphone cables carry both the audio signal and an inverted version of the signal too, in order to reduce noise that could be introduced along the signal wire. The result of this is that if a balanced microphone cable is wired up wrong, then the polarity of

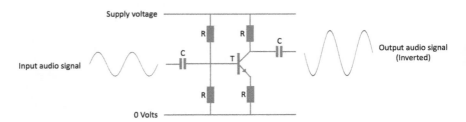

Figure 14.10 Single-stage transistor amplifier circuit, giving an inverted signal output (C = capacitor; R = resistor; T = transistor).

the audio signal can be flipped or inverted. This is not a good thing, because if done accidentally, then mixing a signal with the opposite polarity version of the same signal will cause cancellation and result in no signal at all!

So we can see the likeness between adding two identical signals, one with the polarity flipped, and adding two signals with a 180-degree phase shift caused by a time delay, but they are fundamentally different. Unfortunately, over the years, people started to replace the term *polarity switching* with the phrase *phase inverting*, which is a bit of an unfortunate term – firstly, because *polarity switching* is completely the correct term – the polarity has been switched – whereas *phase inversion* doesn't make literal sense, since phase can only be shifted, not inverted. Polarity switching doesn't really have any-thing practical to do with phase, though mathematically these concepts can be shown to be related in a complex way when applied to pure sine waves.

There is a second occasion where signal inversion might be observed, and that is with microphones which point in different directions or orientations. Now, back to the issues and benefits of polarity switching; the reason why mixing desks have the polarity switch on them is to allow the recording engineer to make an informed decision on whether a sonic improvement can be made to the sound of the recording by switching the polarity. The example of a drumhead vibrating is a good one because it is theoretically possible to record a drumhead on both the top and the underside, with the micro-phones both pointing in opposite directions, as shown in Figure 14.11a. When one microphone records a positive signal, the other records a very similar negative signal. In this case, mixing the two signals together will cause serious cancellation. Switching the polarity of one microphone signal will allow the two signals to be mixed or added together without cancella-tion occurring. In reality, the perfect-storm set-up as shown in Figure 14.11a very rarely occurs, and it's usually the case that a snare drum is recorded with microphones positioned close to the batter and the underside of the resonant drumheads, as shown in Figure 14.11b.

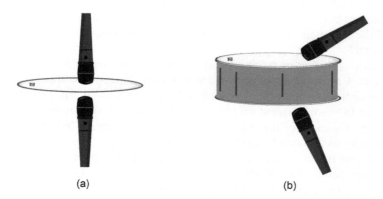

(a)   (b)

Figure 14.11 Microphones positioned with different orientation on drumheads and drums.

This scenario is much more complex, firstly because the microphones are not perfectly aligned in opposite orientation (they more likely have a 90-degree physical offset than a 180-degree offset). Secondly, there is a physical displacement between the two microphones also, given that the top head is struck with a drumstick, and the batter head is placed around 15 cm away, depending on the depth of the drum. In this instance, it's impossible to know if the polarity switch will improve or worsen the sound of the two microphones mixed together, you just have to try it and decide by listening for yourself! It's useful to employ a similar approach when using more than one microphone on a kick drum too, even if they are both pointing in the same direction, you may or may not hear an improvement by switching the polarity of one of the signals. If you are getting a poor sound from your drum kit when set up to record, signal polarity might just help explain what is going wrong.

TRY FOR YOURSELF: EXPLORING POLARITY AND POLA-RITY SWITCHING

Set up the scenario shown in Figure 14.11a – this can be done by removing a resonant drumhead and positioning two identical microphones at identical distances above and below the drumhead. Either with a mixing desk or in a DAW after you have made a recording, compare the sound of the two microphones mixed together with and without polarity switching. What is the difference in the sound? Does cancellation occur and is this reduced with the polarity switch activated? Take a look at the signal waveforms, can you see one signal has a different polarity to the other?

Now consider the more realistic scenario with a two-headed snare drum and two microphones positioned at an angle to the drumhead (Figure 14.11b). Repeat the exercise and listen to how the sound changes between using signal inversion and no signal inversion on one of the channels. Is there much difference, and does this improve the sound?

Add another microphone facing the batter head, but distanced about 1 m away. Does this cause any cancellation when mixed with one or both of the other two microphones? Does switching the polarity of the more distant mic improve the sound? Experiment with different distances of the third mic too, to see how much help the signal polarity switch can be in different scenarios.

If you make a quick DAW recording of some snare hits with this set-up, you'll also be able to see the polarity and time delay detail of each microphone's captured waveform, similar to that shown in Figure 14.5.

Notes

1 Interview with producer/engineer Sylvia Massy conducted on 30/08/2020.
2 A 3-ms time delay will cause a sine wave of 6-ms period to have a 180-degree phase offset. A 6-ms period sine wave can occur 1/0.006 times per second, which is equivalent to 166.7 Hz.
3 Many DAW spectrum analyser plug-ins do not have sufficient frequency accuracy to show the full effect of comb filtering. Professional metering plug-ins, such as the Nugen Visualizer, allow high-resolution frequency analysis, which can be valuable for identifying phase and other coherency issues when working with audio signals. More details on the Nugen Visualizer are available at https://nugenaudio.com/visualizer/ [accessed 01/08/2020].

15 Microphone techniques for recording drums

In previous chapters, we've discussed microphone and equipment types, room acoustics, and the non-trivial topic of phase, which collectively give the ideal foundation for covering microphone placement techniques in more detail. The theory and expert technique of recording music has its pinnacle in the skill of choosing what microphones to put where in a performance space. Given the number of instruments and microphones used in a drum recording session, these skills are challenged to their furthest limits when tracking drums. There are many options and many different approaches which are preferred by renowned studio engineers and music producers. Indeed, you'll find often that two brilliant established professionals will rarely do things the same way. This is because there are no hard-fixed rules in the creative act of recording and perhaps the most important aspects are experience and reflective practice. It's impossible to test every microphone that is available in combination with every microphone technique, so experienced engineers learn a number of methods which suit their style and the types of projects they work on, and the types of equipment they have available, allowing them to perfect their own approach. Another engineer might do things in a completely different way, yet still have a perfected approach which yields comparable results. The key to developing expertise here is to experiment with many different microphones and techniques initially, always reflecting critically on the results and gradually developing towards a personal approach. Eventually you can focus on using a smaller variance of microphone types and an appreciation of the microphone techniques which give you the best results in relation to your personal objectives or those of the project.

15.1 Microphone placement approaches

There are generally three broad microphone placement techniques for drums, and many recording engineers use a combination of these approaches to capture all sonic characteristics and nuances of the drum kit, whilst also allowing necessary production options during mixdown. It's possible to differentiate these three recording techniques by the distance of the microphone from the drum kit, i.e. close, near, and far, though in general it's more valuable to term each technique more descriptively based on the purpose of each microphone. *Spot microphones* are those which are placed very close to a specific drum or cymbal on the drum kit, in order to pick up a detailed and focused recording of individual instruments in the kit. *Stereo microphones* require at least two microphone capsules and are generally placed near but a few feet away from the drum kit, in order to pick up an authentic representation of the set-up, capturing a good balance of each drum in the kit and with a natural representation of the drums and cymbals as they are physically positioned from left to right on the kit. Often these are referred to as *overhead microphones* because they are usually placed above or at the head height of the drummer. *Room microphones* are place more distant and further away (perhaps 2–5 m away) in order to capture sonic characteristic of the drum kit in combination with the performance space that is being used for the recording. With these microphone locations in mind, there are therefore a number of combinational approaches for recording a drum kit, as given in Table 15.1, with some summarised advantages and disadvantages of each approach.

It's clear from Table 15.1 that each drum recording approach has its own merits and challenges, and so all of those included should be considered equally valid techniques. Using more microphones does not always yield bigger and better results, and likely bring more challenges with respect to phase coherency and authenticity in the mixed recordings. Moreover, the different techniques are more suited to different projects, different recording scenarios, equipment availability, budgets, and different music genres, and it is valuable for a recording engineer to be experienced, confident, and capable of recording a viable drum sound with each of the approaches described in Table 15.1. There are also many different approaches for each technique given, incorporating the choice of microphone and the exact placement of microphones dependent on their purpose and role in the recording set-up. The following sections discuss aspects of each approach in more detail.

15.2 Stereo recording for drums

15.2.1 Defining the stereo field

There are two important concepts to consider when choosing a stereo recording technique for a project, both relating to the physical positioning of sounds from left to right, which we are able to hear and distinguish between with our two ears. Firstly, we are able to identify direction and positioning of sounds, so stereo techniques allow us to accurately playback the positioning of sounds as they were recorded. Quite simply, if we use a stereo microphone technique to record two vocalists singing on different sides of a stage, then when we play back the recording in stereo (i.e. with two loudspeakers), we can hear one singer more from the left and one singer more from the

Table 15.1 Drum recording approaches

N	Stereo	Spot/Close	Room	Advantages	Disadvantages
2–4	Stereo overhead microphones	None	Optional	Simple and quick set-up Natural sounding recording Low risk of phase coherence issues	Less definition to individual drum sounds Limited stereo width on playback Limited control of volume balance of drums and cymbals in mixdown
3–6	Stereo overhead microphones	Kick drum mic Optional snare drum mic	Optional	Simple and quick set-up Good definition to kick and snare drum sounds	Phase coherency between microphones needs evaluation during set-up Limited stereo width on playback Limited control of volume balance of drums and cymbals in mixdown
6–12	Stereo overhead microphones	One or more microphones for each of the drums in the kit	Optional	Excellent definition to all drums in the kit Greater stereo width of toms through panning during mixdown	Phase coherency between microphones needs critical evaluation during set-up Limited stereo width of cymbals on playback
8–20	Stereo overhead microphones	One or more microphones on each of the drums in the kit Microphones positioned close to each group of cymbals	Optional	Excellent definition to all drums in the kit Greater stereo width of toms through panning during mixdown Greater stereo width of cymbals through panning during mixdown More control of volume balance of drums and cymbals in mixdown Allows a hyper-real drum sound after mixdown	Complex and time consuming set-up Phase coherency between microphones needs critical evaluation during set-up Large data space required for recording More complex and time-consuming editing and mixing

N = total number of microphones used.

right, to varying degrees of authenticity depending on the microphone technique used. A second stereo consideration comes from the reality of being in a space and hearing a music performance from different positions within that space. For example, a piano has the bass keys on the left-hand side and the treble keys on the right-hand side, and if you sit and play at a piano, you will hear a distinct left-to-right positioning of bass and treble sounds as you play. However, if you sit in the audience and listen to someone playing the piano from a distance of, say 5 m or more, the bass and treble sounds from the piano become completely merged and impossible to hear the difference between their left and right positioning on the keyboard. In this instance, the stereo sound becomes much more respective of the environment and the room in which the piano is positioned. We can therefore also consider stereo sound as being a blend of both the direct sound we hear from the instrument as it travels straight to our ears and all the reflections and reverberation from the space and the room, giving us sonic characteristics from the side and back walls, and the ceiling and floor too. So we have a choice when recording and mixing stereo sounds, do we want the stereo to represent the positioning of the instruments and the positioning of sounds within those instruments, or do we want stereo to be used to capture the relationship between the instrument and the room in which the instrument is positioned? Of course, we can have our cake and eat it, and have a blend of both scenarios, but the choices of stereo microphone techniques used depend very much on these two concepts and your preference for the nuances of the final sound that will be played back after recording and mixing has been conducted.

Additionally, our ears recognise the source positioning of sounds by assessing at least four sonic attributes, and different stereo microphone techniques take advantage of these attributes in different ways. Firstly, we can identify the positioning of sound based on the volume of the sound when it reaches our ears. If the sound reaching the left ear is slightly louder than the sound reaching our right ear, we know that the sound must be coming from the left-hand side with respect to where we are situated. Secondly, our ears are capable of identifying if a sound was heard slightly before or after in each ear; sound coming from directly in front reaches both our ears at exactly the same time. Sound coming from the left reaches the left ear a fraction of a second before it reaches our right ear (some might call this a phase difference!), so our brain is able to work out where the sound is coming from. The third attribute relates to the physical size and shape of our head; sound passes through the head as vibration (in the same way it passes through walls and materials), but the sound changes somewhat during that vibration. Moreover, sounds hitting the closest ear are directly from the sound source, whereas those hitting the farthest ear have been modified by the mass and density of our head. This results in a slightly duller sound on one side, and our brains are able to convert this difference into information related to where the sound has come from. Furthermore, our brains are also very in tune with the balance of direct and reverberated sound; we are intrinsically able to identify if the sound on our left-hand side has more reflection or reverberation than the sound from the right-hand side, and this helps identify if a sound is closer to one side of a room and has been reflected or not when it reached us. Similarly, we identify how far away sounds are based on this balance too, since sounds from further away experience more reflections and reverberations as they reach our ears than closer sounds do.

We use the term *stereo field* to represent the positioning of instruments and sonic attributes from left to right. Often a creative objective when recording multiple instruments in stereo is accurate localization in the sound playback. This means that instruments in the centre of the group should ideally be reproduced to give the impression that they are coming from midway between the two speakers (which we call a *phantom centre image*). Instruments at the sides of the group are heard to come from each side. It's actually very difficult to record in stereo and maintain the natural stereo field, and sometimes we want to manipulate the stereo field after recording to make it sound wider or narrower too. To give an example, Figure 15.1 shows four different stereo images on playback of a drum recording through loudspeakers. When listening to a drum kit recording, the natural stereo

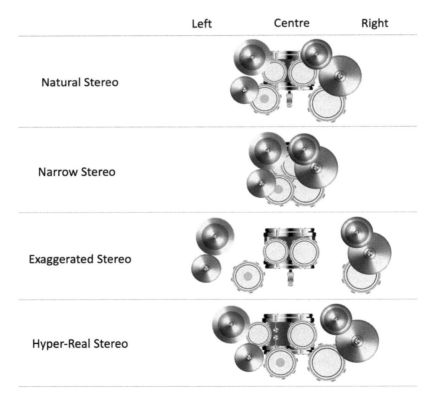

Figure 15.1 Potential stereo images of the kit set-up on playback of the recording.

image sounds like a drum kit as if you were playing it yourself or stood just in front; the kick drum is in the centre, the snare slightly to one side, toms spaced around and just off the centre, and with the cymbals perhaps a little wider to the left and right too.

Whilst it's valuable to be able to capture an accurate stereo image of a sound source, with drum recoding and production we don't always want a completely accurate stereo image in the playback. For some projects and music genres, it may be more appropriate to have a narrower stereo image for the drums, to allow other instruments to occupy the extreme left and right positions in the mixdown. For other projects, we might prefer a very exaggerated stereo image, with kick and snare drums occupying the centre and cymbals and toms all panned hard to the left and right, giving impact and extreme positioning between the drum sounds when we listen back. Additionally, a decision on how far back you would like the listener to feel when hearing the recordings can be considered, because drums further away give a narrower stereo image in reality for the listener, whereas someone sitting directly in front or behind the drums experiences a more exaggerated stereo image of the instrument. The *hyper-real* example is fairly common in drum production and can sometimes be created by repositioning drums in the stereo field during mixing. If you have recorded with spot microphones on each drum and cymbal, you can reposition things somewhat afterwards, for example placing the snare directly in the centre, moving the toms far and wide from left to right, pulling the hi-hit more central and pushing the ride and crash cymbals to the extreme left and right positions. There is often a trade-off with the hyper-real approach in that inauthentic positioning will most likely introduce phase coherency issues, but the producer and mix engineer can decide for themselves if the benefits outweigh the deficiencies this brings. One challenge with drum recording is to have an understanding of what the final mix should sound like during the recording stage, if you want an exaggerated stereo field, yet your recordings only allow a narrow image, then you will be limited in how closely you can achieve your creative objective in the mix session.

15.2.2 Spaced pair technique

The *spaced pair* method (also called A-B) involves two identical microphones placed a few feet apart and aimed straight ahead (as shown in Figure 15.2). The mics can have any polar pattern, but omnidirectional is most popular for this method. The greater the spacing between mics, the greater the stereo spread.

The spaced pair records the inherent delay in sound reaching one mic before the other, which is relative to the instrument positions in the performance space. During playback, the brain interprets these time differences into corresponding image locations. A delay of 1.2 ms is scientifically enough to shift an image all the way to one playback speaker; this value is

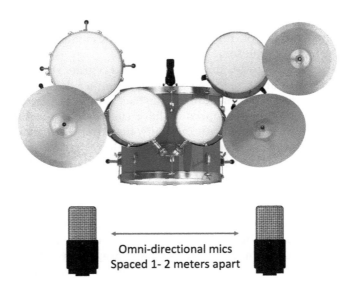

Omni-directional mics
Spaced 1- 2 meters apart

Figure 15.2 Spaced pair or A-B microphone technique.

based on the width of the human head and the fact that sound from a directly left source will reach the right ear 1.2 ms after it reaches the left.

We can use this fact when setting up mic positions. Suppose we want to hear the right side of the drum kit from the right speaker when playing back. The sound from the right side of the kit must reach the right mic about 1.2 ms before it reaches the left mic. We know that sound travels at approximately 30 cm in 1 ms, so the right microphone needs to be about 30 cm closer than the left mic. Depending on how high or how far away from the kit you are, a bit of quick trigonometry shows that the microphones need to be spaced about 1 m apart. If the spacing between mics is exaggerated, say 2 m, then instruments that are slightly off centre produce delays between channels that are greater than 1.2 ms. This places their images at the left or right speaker and can create the exaggerated effect shown in Figure 15.1. The time delay theory holds true, but in reality the spaced pair method tends to make off-centre images unfocused or hard to localize or pinpoint. This is because stereo images produced solely by time differences are not very sharp, our ears and brains use much more information than time difference alone. The spaced pair technique is therefore a good choice to achieve sonic images that are diffused or blended, instead of sharply focused. Unfortunately, with spaced microphones, given the inherent time delay characteristics, if both mic channels are mixed to mono, phase cancellations can be found at some frequencies. Spaced mics, however, give a strong sense of ambience, in which the reverb seems to surround the instruments and sometimes the listener too, owing to the diffuse reverberations that each microphone picks up. Another advantage of the spaced pair technique is that you can use omnidirectional mics, which tend to have a more accurate and authentic frequency response than directional microphones.

15.2.3 X-Y technique

We often use the term *coincident pair* when referring to two microphones that are positioned in the same place but pointing in different directions. One of the simplest and most popular coincident pair techniques is the *X-Y pair*, as shown in Figure 15.3. For balanced stereo reproduction, the two mics should be the same model and preferably matched (meaning that the manufacturer has taken extra care to ensure their characteristics are identical). With the X-Y technique, it's important to use microphones that have directional polar patterns: cardioid, hyper-cardioid, or super-cardioid. The most important benefit of all coincident pair techniques is that they have very little phase incoherence and are almost perfectly mono compatible, because the two microphone capsules are equidistant from all performance sounds.

The angle between the two capsules in an X-Y pair is usually set between 90 and 120 degrees, where the wider the angle, the broader the stereo spread. The stereo image is reproduced because the directional mic pointing left picks up sound straight ahead, but rejects that from the side (i.e. the right), and vice versa for the right mic, so the stereo image is based on the volume or sound intensity level recorded on each channel. If the angle is too wide, or the mics are too directional, instruments in the centre can have a reduced level and less sharpness in playback, because both microphones are *off-axis*

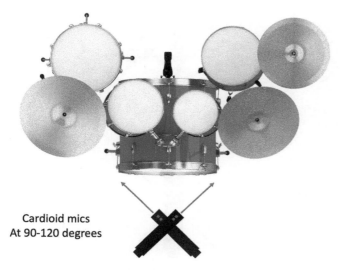

Cardioid mics
At 90-120 degrees

Figure 15.3 Coincident pair (X-Y) stereo recording technique.

to the centre. There may also be sound coloration due to the fact that off-axis sound constitutes a significant proportion of the recording. Some mics have poor off-axis frequency response, so this should be taken into account, particularly given that the kick and snare sounds are generally positioned to the centre of a drum kit when recording.

Naturally, the angle of the mics also affects the capture of reverberation and other ambient elements, though in general the X-Y technique (using directional microphones) captures less natural ambience than other techniques. An X-Y pair angled 90 degrees apart, for instance, can cause a build-up of room reverberation in the phantom centre, whereas wider angles spread the reverberation more across the stereo field.

15.2.4 Blumlein pair technique

The *Blumlein pair* configuration, shown in Figure 15.4, is very similar to the X-Y technique, but uses two bidirectional (figure-of-8) microphones whose capsules are angled 90 degrees apart. Invented by audio pioneer Alan Blumlein (who is credited with inventing the concept of stereo itself), the Blumlein pair creates a four-sided polar pattern that gets summed to two channels. The left channel is made up of front-left and rear-right signals, and the right channel consists of front-right and rear-left signals.

The Blumlein pair is a traditional coincident pair technique that has all the same benefits with phase coherency as the X-Y technique. The two main differences are that, firstly, the Blumlein pair takes advantage specifically of

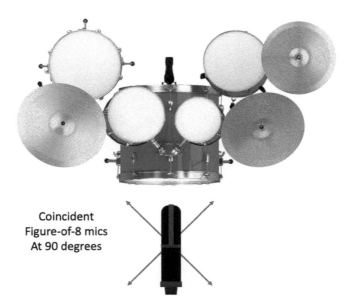

Figure 15.4 Blumlein pair microphone technique.

ribbon microphones, which give a warm yet detailed sound. Secondly, the Blumlein microphones' figure-of-8 patterns cause more rear facing sound to be picked up, meaning more room sound is captured, which may be an advantage or disadvantage dependent on the acoustics of the recording space and the production objectives. Some microphones such as the AEA R88 and the Royer SF-12 are built as stereo Blumlein microphones, with two perpendicular ribbon diaphragms inside and giving stereo signal outputs, similar to the simple microphone shown diagrammatically in Figure 15.4.

15.2.5 ORTF technique

The *ORTF* method is a *near-coincident pair* method. This technique was developed by French radio station Office de Radiodiffusion Television Française (hence the technique is named *ORTF*) and uses two cardioid microphones spaced by around 15–30 cm (roughly the same as ear spacing) and angled outwards at around 80–120 degrees, as shown in Figure 15.5. This technique often gives a greater sense of space than standard coincident techniques due to the microphones being ear spaced and thus capturing time delay information as well as sound intensity information. As with the X-Y technique, the greater the angle or spacing between mics, the greater the stereo spread.

The ORTF technique is not perfectly mono compatible, but phase coherency issues are usually quite minor, given the near-coincident set-up. The technique has good stereo differentiation between left and right sounds, because of the directional microphones used and the additional spacing and time difference in the recorded stereo sound. As with all coincident

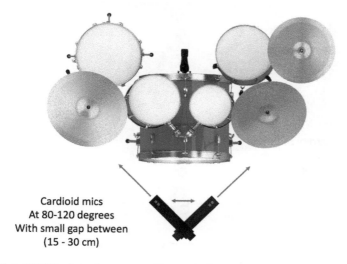

Cardioid mics
At 80-120 degrees
With small gap between
(15 - 30 cm)

Figure 15.5 ORTF microphone recording technique.

techniques, the directional microphones allow good rejection of ambient or reverberant sound. The spacing between the microphones also allows shallower angles to be used than with the X-Y technique, meaning that it is possible to achieve more focused centre-positioned sounds too.

15.2.6 Mid-side technique

The *middle-side, mid-side*, or *M-S* stereo pair is a coincident technique that involves an extra layer of complexity, but can yield excellent results. The mid-side method, as shown in Figure 15.6, uses one *mid* microphone (typically a directional mic with a cardioid pattern) that faces the sound source and one bidirectional (figure-of-8) *side* microphone which is directed perpendicular to the mid mic.

One side of the figure-of-8 is regarded as the positive side (the front) and one side is regarded as the negative side (the back). In Figure 14.6, the orientation of the front is pointing to the left, as shown by the "+ve" sign.

The mid-side technique doesn't immediately create a stereo audio signal; in order to create an actual stereo image, we need to perform a little post-processing on the recorded audio. It is possible to convert mid and side channels to stereo left and right, and to convert left and right stereo signals to mid and side with a simple mathematical equation and a set-up that can

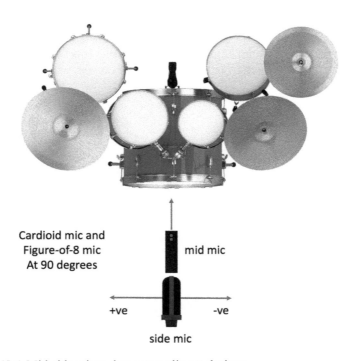

Figure 15.6 Mid-side microphone recording technique.

be implemented in a digital audio workstation (DAW). Consider the mid microphone to record sound with components from both the left and the right, and the side microphone records sounds from the left as positive signals and sound from the right as negative signals. If we add (sum) together the mid (M) and side (S) microphone signals, the two left components add together, but the right component of the side mic cancels out the right component of the mid mic. So, the left-hand side of the stereo field (L) is defined by the equation

$$L = M + S$$

If we subtract the side microphone signal (or invert it then sum) from the mid microphone signal, we get the opposite effect with cancellation on the left, so the right-hand side of the stereo field (R) is defined by the equation

$$R = M - S$$

This technique is often referred to as the *sum-and-difference* method, and it's possible to reverse the formulas to calculate mid and side components from left and right data too.

Once a mid-side recording has been captured, it's necessary to mix the two signals to give true left and true right channels as follows:

- Mix the mid microphone with the side microphone to give LEFT (i.e. $M + S$).
- Mix the mid microphone with a polarity inverted copy of the side microphone to give RIGHT (i.e. $M - S$).

There are a few different ways to achieve M-S to L-R decoding and encoding in a DAW, and plug-ins exist for this too. It's possible to achieve with some simple signal flow set-up in any DAW as follows (and as shown in Figure 15.7):

- Import the mid and side audio into the DAW arrange screen.
- Copy the side audio to a third track and insert a polarity invert plug-in on the channel.
- Hard pan the side audio to the left and hard pan the inverted side audio to the right.

With a mixer or DAW, it's possible to adjust the pan and volume settings to decide how much cancellation occurs in the conversion, which in effect allows us to decide how wide we want the stereo image to be – even after the recording has taken place! If we pan the two side channels (the original and the inverted) to the centre, they simply cancel each other out, leaving just the mid mic, so it's possible to choose how much stereo effect you want. In fact, wherever the pan settings are, if we bounce down to mono, the side

Figure 15.7 Mid-side decode method in Logic Pro. (Gain plug-in is used to achieve polarity inversion.)

channels will always cancel, leaving just the mid mic. Mid-side is therefore perfectly mono-compatible and allows some valuable manipulation of the stereo image in post-production. If recording drums with mid-side, the mid microphone captures a strong overall sound of the kit and particularly the centre components of the kick and snare. The left and right extremes of the kit are less precise in terms of placement across the stereo field, but a strong and impactful ambience of the room and the stereo field can be achieved, whilst maintaining a good accurate representation of the drum kit as heard in the recording space.

TRY FOR YOURSELF: MID-SIDE CONVERSION AND STEREO COMPARISON

You can try the mid-side processing as shown in Figure 15.7. If you don't have any mid-side recordings available, then you can download an example orchestra recording using each of the mid-side, spaced pair, and Blumlein techniques, produced by Rob Toulson.[1]

Bring into a DAW session the mid and side microphone recordings, then duplicate the side recording and invert one version of the side. Now pan the side signal to the left and the side-inverted signal to the right. Start with the volume of the two side channels turned completely down and listen to just the mid microphone recording. Now bring up the side and side-inverted channels in volume and listen to how the stereo image opens up with much more character, width and detail.

It's also valuable to compare different stereo recording techniques applied to a single instrument. So if you get chance to record a drum kit with the mid-side technique and also one or two of the other techniques described in this chapter at the same time, then this serves as an excellent learning experience to practically evaluate the qualities of each approach.

15.2.7 Baffled omnidirectional pair technique

The *baffled omnidirectional pair* method uses two omnidirectional mics, usually ear-spaced, and separated by a hard or padded baffle. This technique is designed to replicate two ears on either side of a human head, as shown in Figure 15.8. To create stereo, it uses time differences at low frequencies and level differences at high frequencies. The baffle creates a sound shadow (reduced high frequencies) at the mic farthest from the source. Between the two channels, there are also spectral differences (differences in frequency response) given that the baffle absorbs some frequencies more than others, in the same way that a human head does.

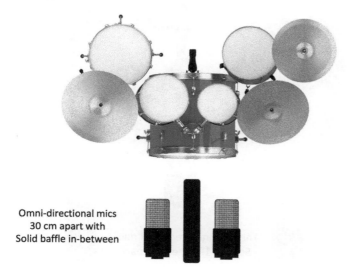

Omni-directional mics
30 cm apart with
Solid baffle in-between

Figure 15.8 Baffled omnidirectional pair microphone technique.

The baffled omnidirectional pair therefore uses level, time, and spectral differences to produce the stereo effect, resulting in sharp images and accurate stereo spread. This method has a good low-frequency response owing to minimal phase cancellation issues, especially at low frequencies. The Neuman KU 100 Dummy Head microphone is an example of a baffled stereo pair mic that can give great results as a drum overhead; it quite literally aims to mimic the human head and ears in design and microphone response.

15.2.8 Decca tree technique

Spaced omnidirectional microphones have an accurate and spacious sound, and work well for the spaced pair technique discussed earlier. The wider the separation, the more spacious, but if they get too wide, then instruments at the centre of the sound stage become less focused. Too wide can also result in issues with side wall reverb and reflections that are a little too strong. The British classical music record label Decca devised a microphone technique that solves some of these problems by placing a third mic in between and in front of the spaced pair, as shown in Figure 15.9. The output of the centre microphone is sent to both the left and right channels, and its volume can be set to give the desired balance of centre sound.

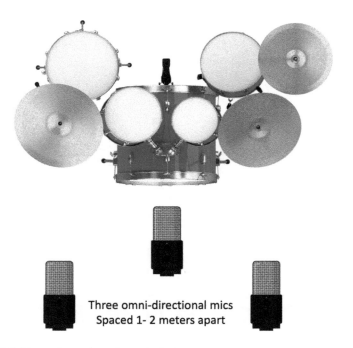

Three omni-directional mics
Spaced 1- 2 meters apart

Figure 15.9 Decca tree microphone technique.

This configuration, known as a *Decca tree*, has been used by Decca on many of their orchestral recordings for a number of years. This can be a good technique for drums too where a wide omnidirectional spacing is desired to get a strong stereo image, but also to allow a focused and powerful centre component (i.e. kick and snare) to be captured too.

15.2.9 Left-right-centre technique

The *left-right-centre (LRC)* method is very similar to the Decca tree idea, but uses a more coincident approach. Instead of using three spaced omnidirectional microphones, it uses a near coincident X-Y pair (similar to the ORTF pair) with a third directional microphone positioned towards the centre, as shown in Figure 15.10. The result is a strong and clear representation of both sides and the centre of the sound source too, with few phase coherence issues. These attributes make LRC a good choice for drums, particularly if only three microphones are available for recording and adding further spot microphones is not an option.

15.3 Using spot microphone techniques

Whilst drum overheads capture an authentic representation of a drum kit in a performance space, modern music productions tend to incorporate more hyper-real techniques to achieve greater impact and intensity with drums. It is not uncommon for drums to be heavily compressed, adding greater

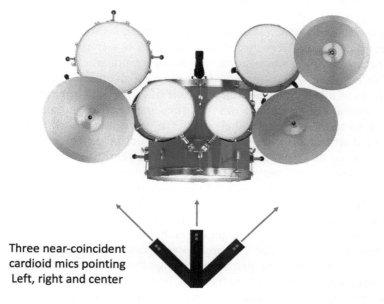

Three near-coincident cardioid mics pointing Left, right and center

Figure 15.10 Left-right-centre recording technique.

consistency and definition, shaped to give fast attack times and controlled decays to each drum sound, and hard panned to use the full width of the stereo field. Much of this hyper-real production is not possible with capturing sound from overhead microphones alone, which contain elements of all drums in all microphones. If you want to make the snare stand out to an extreme level in the mix, this is very difficult to achieve if you have only used overhead microphones. As a result of the desire for hyper-real drums in popular music productions, using close or spot microphone placements on some or all drums has become very common nowadays.

Often it is valuable to incorporate close microphones on the snare and kick drum, since these drive the groove on most popular songs. In combination with using stereo overheads to capture the cymbals and toms, close microphones on the snare and kick drums can give a solid sound to any production. With this in mind, it is very possible to capture an excellent drum sound using just four microphones, as mentioned previously in Table 15.1. There are many different techniques for close microphones, with subtle variations for each type of drum.

15.3.1 Kick drum microphone technique

The purpose of close microphones is to capture a sound that can be effectively used in the mix to achieve the musical objectives, which usually relate to tightness, clarity, and power for drums. In this case, it is valuable to capture all the sonic nuances of a kick drum and, with some knowledge of drum acoustics, it's possible to make good informed choices. Firstly, the kick drum gets its power from the strong fundamental F0 frequency which is excited by the beater in the centre of the kick's batter drumhead. To record this, a specialist low-frequency microphone is required which has a good frequency response extending down to 50 Hz or lower. Positioning a dynamic kick drum mic inside the kick drum itself allows the fundamental tone of the kick to be captured, and a good amount of isolation from the other drums and cymbals is achieved owing to the kick drum's shell itself. To get inside a kick drum, many resonant drumheads are supplied with a port hole, which allows a microphone to be positioned inside whilst also retaining the qualities that the resonant drumhead gives to the drum sound. It's also valuable to experiment with the angle and positioning of the mic inside the kick drum, since this can have a significant effect on the tone and phase coherency of the recording. Placing the internal mic closer to the batter head can also capture some useful impact of the beater hitting the drumhead too. It's noted that an easier solution to recording inside the kick drum is to remove the resonant drumhead altogether; however, the kick drum becomes a very different instrument when it has only one drumhead. The resonant drumhead not only provides an essential component of the acoustic system of the kick drum but also gives a pressure response to the kick pedal, owing to the air mass trapped inside the drum, which many drummers use to their advantage during a performance. In some cases, engineers have even wired a kick drum mic into the drum itself in order to achieve an internal close mic recording, without having to remove or cut a hole in the resonant drumhead.[2]

It is not uncommon, however, for the internal kick drum mic to lack some sparkle and clarity, since it is focused purely on capturing the fundamental low energy of the drum. If this is the case, then positioning a good quality and robust condenser microphone on the outside, facing the resonant drumhead, can capture the clarity that is missing from an internal mic. Again, it's valuable to experiment with the distance from the drum, since a few centimetres can change the captured sound significantly, especially when blended with the other microphones. Additionally, or instead, it might be worthwhile to place a cardioid microphone facing towards the kick pedal beater and the position where it contacts with the drumhead, in order to capture a more crisp attack profile of the kick sound. Figure 15.11 shows three different microphone positions for kick drum recording.

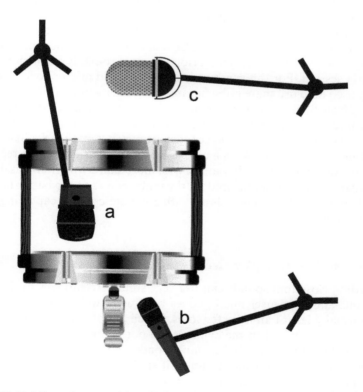

Figure 15.11 Microphone positions for kick drum recording: (a) inside; (b) beater position; (c) external.

15.3.2 Snare drum microphone technique

The snare drum is often the drum which producers and recording engineers spend most time evaluating and crafting in the recording and mix sessions. Whilst much can be done with tuning and setting the right pitch of the snare, microphone choice and positioning can make a huge difference too. It's an interesting observation that while three snare drums might sound the same in a room, they will usually sound very different when recorded using a close dynamic microphone. Likewise, a single snare drum can sound very different by changing the microphone type or adjusting the mic position while recording. For example, the angle of the microphone can dictate the balance of fundamental and overtone frequencies that are captured, since pointing towards the edge will capture more significant overtones of the drum and pointing towards the centre will focus more on the drum's fundamental frequency, as we have seen previously with the iDrumTune analysis of drumhead frequencies. Ergonomics are an issue here too, since the close snare microphone needs to not interfere with the drummer's performance, and sometimes getting the perfect position can be tricky to achieve. Placing the snare mic towards the edge of the drum's batter head, pointing down at an angle and inwards towards the centre, usually gives a good balanced sound of the drumhead's fundamental and overtone frequencies, and the timbre of the snare shell itself. Spillage is also a consideration, since the snare and hi-hat are so close together, yet an isolated snare sound is much more useful in a mix session. For this reason, dynamic cardioid microphones are commonly used for close snare mics, because they offer good rejection of sounds coming from behind the microphone and the nature of a dynamic mic is less ideal for recording bright sounding cymbals anyway. It is possible to record a great snare sound with a robust condenser mic, but often the spillage of the hi-hat makes it less useable in a mix session. It can also be worth recording the underside of the snare, if a sound with more snare-wire timbre is desired. A similar technique can be used on the underside as to the batter head, except there is usually no major issue with ergonomics when capturing sound from the underside.

15.3.3 Tom drum microphone technique

Recording close tom sounds can use very similar techniques to the snare. Dynamic and condenser microphones both work well on toms, and spillage is usually less of an issue for toms, which are played less regularly and often at a moment when the cymbals are not being hit, allowing them to be manipulated more easily in a mix. Condenser microphones can give an excellent clarity and warmth to toms, if spillage from cymbals is not a big issue. Producer/engineer Mike Exeter uses a novel approach for capturing tom sounds, which allows the timbre of the resonant drumhead to also be captured, whilst giving excellent isolation and reducing spillage from other drums and cymbals in the kit. The technique involves placing close dynamic microphones on the top and bottom drumheads of each tom and linking the two audio signals together. Firstly, the underside microphone has its polarity switched, since the resonant drumhead microphone is oriented to pick up sound pressure in the opposite direction to the batter drumhead, then the two microphone signals are mixed together to give a complete representation of the drum sound. The approach has the added benefit of *common-mode rejection*, which means that anything common to both microphones (i.e. the other drums and cymbals in the kit) are forced to cancel out, since one of the microphone signals is inverted. The process of inverting and mixing can be conducted either in a DAW or in the microphone XLR (*external line return*) cables themselves. Mike Exeter explains

> The technique involves having a single "y-lead" with 2 female XLRs at the mic ends to a single male XLR to the preamp. One of the female XLRs is wired with the positive and negative pins reversed. This therefore takes the out of phase signals from the top and bottom tom mics and puts them back in phase at source before hitting the mic preamp. The benefits are that you only use one recording channel and commit the sound, plus the ambience/spill is also phase reversed between the mics, which reduces the overall spill and allows for a much punchier and clearer tom hit. I use the same mics on top and bottom (Sennheiser E604) so the levels are ok to be the same, but I have sometimes used variable in-line pads to reduce the bottom mic levels to tweak the balance between top and bottom.[3]

Whilst it is not essential to use the two-microphone technique on tom drums – and great tom sounds can be captured with a single mic only per drum – this is a clear example of an esteemed recording engineer who has perfected their own approach to recording based on their own experience and critical evaluations of different techniques.

15.3.4 Close cymbal microphone technique

When attempting to create a hyper-real mix that might involve hard panning of cymbals and toms, the overhead microphones alone will rarely allow for the most extreme positioning of sounds. If this is desired, then it can also be useful to incorporate close microphones on cymbals or cymbal groups. It is often valuable to capture a close hi-hat to be tamed and balanced in the mix if necessary; it's not uncommon for a simple recording to have an overly loud hi-hat sound which diminishes the impact of the other drums in the mix. Capturing close microphone recordings of the hi-hat can allow specific attention to be paid, for example lowering the relative volume, filtering some high and low frequencies, and moving the pan position wider or more central as desired.

Figure 15.12 Close microphones positioned to capture cymbals: (a) hi-hat; (b) crash; (c) ride and crash.

It can also be worth using close microphones for the other left- and right-positioned cymbals, as shown in Figure 15.12. In particular, the ride cymbal is often performed with less force than other cymbals, and a fairly close microphone positioned towards the ride can allow this to be enhanced and made more prominent in the mix. Equally, close microphone recordings of crash cymbals can be used to give wide and impactful sounds across the stereo field, and long cymbal swells can be automated in the mix to move from left to right, if this is an abstract effect that you like. It's not always necessary to have close microphones on all cymbals; for example, a single microphone can be used to capture close sounds of both a ride and a crash cymbal which are positioned close together on one side of the drum kit.

15.4 Room microphones

Room microphones allow a slightly more abstract sound to be recorded which may add some extra sparkle, intrigue, or energy in the final mix. They can also be used to capture qualities of the performance space, if this is preferred to using an artificial reverb in the mix. Room microphones can be mono or stereo and can be positioned anywhere in the room. Some engineers even place room microphones outside the room in an adjacent corridor with the performance space door left open, and others use contact mics placed on the walls. The challenges with using room microphones are that the recorded sounds are generally too ambient or reverberant to be useful on their own, and the distance from the drum kit means that more pre-amp gain needs to be added, and more gain means more likelihood of hiss or noise being captured on a recording. Furthermore, the distance of 2–5 m from the drumkit causes around 5–15 ms of delay to the recorded sound, so some time re-alignment might be required in the mix to avoid audible echo and comb filtering.

Despite these challenges, a room microphone recording can be manipulated, crafted, compressed, distorted, or filtered to give a novel abstract sound, which may add a certain magic or sparkle to a mix if blended in subtly.

15.5 Microphone choices for recording drums

There are hundreds of different microphones available for recording drums, all with different qualities and price points. Nevertheless, nowadays, with good knowledge and experience, it is possible to capture a great drum sound with just a few reasonably priced microphones. Some examples of commonly used microphones for recording stereo overhead and close drum sounds are given in Tables 15.2–15.5. In general, microphones used for overheads are also ideal for use as room microphones and close cymbal microphones too.

Table 15.2 Common microphone choices for use as stereo overheads

Stereo Overhead Microphones	Type	Polar Pattern	Cost (1–5)	Qualities
AKG C414	Large Diaphragm Condenser	Variable	$$	Full frequency range, clarity, reliable sound
Beyerdynamic M160	Ribbon	Cardioid	$$	Ribbon sound with cardioid directionality
Coles 4038	Ribbon	Figure-of-8	$$$	Classic, well-known ribbon drum sound
DPA 4011	Small Diaphragm Condenser	Cardioid	$$$$	Accurate, clear, transient detail
Earthworks SR30	Small Diaphragm Condenser	Hyper-cardioid	$$$	Precision, stereo separation, transient detail
Earthworks TC25	Small Diaphragm Condenser	Omnidirectional	$$	Precision, accurate in all axes, transient detail
Neumann KM184	Small Diaphragm Condenser	Cardioid	$$	Clear, detailed, reliable sound
Neumann U87	Large Diaphragm Condenser	Variable	$$$$	Accurate, full frequency range, detailed sound
Rode NT5	Small Diaphragm Condenser	Cardioid	$	Clear, affordable, reliable sound
Royer SF12	Stereo Ribbon	Blumlein XY	$$$	Stereo mic, classic warm sound, simple and effective

$ to $$$$$ indicate relative cost from low to high in units 1-5, i.e., $ = inexpensive and $$$$$ = expensive.

Table 15.3 Common microphone choices for use as close snare microphones

Snare Mic	Type	Polar Pattern	Cost (1–5)	Qualities
Sennheiser MD421	Dynamic	Cardioid	$$	Robust and clear sound, bright presence
Shure SM57	Dynamic	Hyper-cardioid	$	Classic snare sound, good rejection of cymbals
Neumann U87	Large Diaphragm Condenser	Variable	$$$$	Accurate, full frequency range, detailed sound
Beyerdynamic M201	Dynamic	Hyper-cardioid	$$	Strong, low-frequency characteristic
AKG C414	Large Diaphragm Condenser	Cardioid	$$$	Accurate, full frequency range, detailed sound
AKG C451	Small Diaphragm Condenser	Cardioid	$$	Bright, detailed, full frequency range

Table 15.4 Common microphone choices for use as kick drum microphones

Kick Mic	Type	Polar Pattern	Cost (1–5)	Qualities
AKG D112	Dynamic	Cardioid	$	Low-frequency power, designed for inside kick drum
AKG D12VR	Dynamic	Cardioid	$$	Low-frequency power, warm sound
Neumann U47 FET	Large Diaphragm Condenser	Cardioid	$$$$$	Accurate, warm, low-frequency detail
Avantone Kick	Dynamic	Cardioid	$$	Strong capture of low and sub-bass frequencies
Earthworks SR25	Small Diaphragm Condenser	Hyper-cardioid	$$$	Fast transient response, full frequency range
Electro-Voice RE20	Dynamic	Cardioid	$$$	Tight, focused, robust, full frequency range
Neumann U87	Large Diaphragm Condenser	Variable	$$$$	Accurate, full frequency range, detailed sound
Sennheiser MD421	Dynamic	Cardioid	$$	Robust and clear sound, bright presence

Table 15.5 Common microphone choices for use as close tom microphones

Tom Mics	Type	Polar Pattern	Cost (1–5)	Qualities
AKG C414	Large Diaphragm Condenser	Cardioid	$$$	Accurate, full frequency range, detailed sound
Neumann U87	Large Diaphragm Condenser	Variable	$$$$	Accurate, full frequency range, detailed sound
AKG C451	Small Diaphragm Condenser	Cardioid	$$	Bright, detailed, full frequency range
Sennheiser MD421	Dynamic	Cardioid	$$	Robust and clear sound, bright presence
Shure Beta 57	Dynamic	Hyper-cardioid	$	Robust and clear sound, good rejection of cymbals
Sennheiser E604	Dynamic	Cardioid	$	Clear sound, good rejection of cymbals, clip on

15.6 Developing a personal approach

It's possible to develop new, hybrid, and bespoke techniques from those given in this chapter to develop your own approaches and to learn and understand the benefits of different techniques in different scenarios too. In many cases, the most authentic drum sound is captured by using as few microphones as possible. This concept and approach was adopted and developed further by engineer Glyn Johns, who created his very own technique while recording bands including The Rolling Stones and Led Zeppelin. The Glyn Johns technique uses just three microphones and attempts to capture the best attributes of a drum kit with maximum phase coherency. As shown in Figure 15.13, the method uses two cardioid stereo overheads placed one above and one to the side of the kit, both equally spaced from the snare to allow phase coherency for the snare sound. The third microphone is a close kick drum mic which captures the necessary low-frequency detail of the kit and negates any very low-frequency phase incoherency that might occur with the overheads. It is also possible to add a fourth microphone as a close snare mic, if the desired snare definition is not achieved in the stereo pair.

The Glyn Johns method is a good example of a novel approach that has been experimented with and perfected by an experienced engineer, and one which is still used and elaborated on today. In general, jazz and more classical music desires the drums to be balanced relatively authentically, low, and complimentary in the mix, whereas metal and rock music requires drums that are wide, impactful, and prominent. The key to becoming a proficient drum recording engineer is to understand the musical objectives of the production and to be able to choose a recording set-up that enables the mixdown to achieve the finished sound that is required by the artist, drummer, and music producer.

Figure 15.13 Glyn Johns drum recording method with (a) stereo left, (b) stereo right, and (c) kick drum microphones.

TRY FOR YOURSELF: RESEARCHING PRODUCER TIPS AND TRICKS

Conduct some of your own research on more experimental drum recording techniques, and try these out in a session of your own. For example[4]:

Nirvana producer Butch Vig has often used a long kick drum tunnel to create a controlled and resonant kick drum sound.

Manic Street Preachers producer Dave Eringa has used contact (boundary) microphones attached to the walls of a recording studio to capture an interesting sound that can be blended in with more traditional microphone recordings of a drum kit.

Producer/engineer Sylvia Massy (Tool, Taylor Hawkins) has used a resonator guitar tuned to the key of the song and placed near the kick drum's resonant head to trigger a musical chord every time the kick drum is played.[5]

What are the sonic qualities of the different techniques that you research and try out? Which techniques do you like and will use again? Can you develop some novel or hybrid techniques of your own that work well?

Notes

1 Recording an Orchestra – Six part Video Tutorial by Rob Toulson and comparing different stereo microphone techniques. Videos and full recordings of each stereo microphone technique are available online at https://www.robtoulson.com/orchestrarecording [accessed 01/08/2020].
2 As described by recording engineer and music producer Ken Scott (The Beatles, David Bowie) interviewed by Paul Thompson in his book *Creativity in the Recording Studio: Alternative Takes*, Palgrave Macmillan, 2019, p. 186.
3 Interview with music producer/engineer Mike Exeter, conducted on 21/07/2020.
4 *Kick & Snare Recording Techniques: Tried & Tested Techniques from 50 Top Producers* by Mike Senior, *Sound on Sound Magazine*, June 2008, available online at https://www.soundonsound.com/techniques/kick-snare-recording-techniques [accessed 01/08/2020].
5 *Sylvia Massy Drum Recording – With Hella Comet*, YouTube video by Sylvia Massy, available online at https://youtu.be/xh-B7W4loCI [accessed 01/08/2020].

16 Mixing drums

Balance and dynamics

The most influential tools in mixing can be the most simple. If you have a good quality recording of drums, or any instrument, playing back the raw audio files together should sound great already. If they don't, then you should spend more time developing your recording skills! Needless to say, a great drum recording can be made to sound even more amazing with knowledgeable digital audio workstation (DAW) mixing, using both simple and advanced techniques to make the recordings sound authentic, clear, impactful, and, in many cases – particularly for popular music – better than reality or *hyper-real*.

A key aspect of all music production, and particularly mixing, is listening. There are no exactly right and wrong ways to mix a track, but you may fail to address a particular weakness in the mix if you do not hear something correctly. Things can sound great when really they are not, and as a result, what sounds impactful and exciting in the studio can sound weak, muddy, or dull when you listen somewhere else. If you have a mixing room that uses low-cost speakers or does not have controlled acoustics, then the low-end and bass frequencies can be extremely challenging to mix, and obviously this applies to drums significantly more than many other instruments. You'll find that in an untreated room, the bass will change in volume purely dependent on where you sit. If you are mixing kick drums, toms, or bass guitars, walk around the room, and you'll hear the difference as the bass sounds get stronger and weaker in different locations; sometimes just a few feet can make a huge difference. It's not the purpose of this book to go into studio room acoustics in any depth, but it is valuable to explain the importance of studio acoustics, because if you are not hearing things correctly, then every mix choice you make is potentially compromised or misinformed. If you do have a less-than-perfect mix room, then a really good set of professional studio headphones can help give a good understanding of the bass frequencies in a music track. Spending a few hundred dollars on a set of good-quality headphones is much more cost-effective than spending tens of thousands on room treatment and a professional playback system. Of course, the latter option is always preferred, and you should not underestimate the value of clear and accurate playback systems.

There are many books covering mixing from all angles and incorporating very detailed discussion of room acoustics and mixing a full track. Some well-regarded texts to check out are those by Izhaki,[1] Owsinski,[2] and Savage,[3] for example. In this chapter and Chapter 17, we reflect on the previous topics covered in the book and look specifically at mixing drums, giving a detailed overview of the tools and techniques that can be employed to take a good drum recording towards becoming a professional sounding mix.

16.1 Balance, panning, and bussing

As with ensuring you have a good listening environment, do not underestimate the benefits of having a well-structured mix layout with a clear and organised arrangement of audio *tracks*. If your tracks (or *channels*) are labelled with sensible names, are arranged into well-chosen instrument groups, use carefully considered *gain structure*, and neatly set up *auxiliary sends*, you can keep tight control of your mix as it evolves and invariably becomes more complex. Unstructured mixes can easily result in unnoticed distortions, tracks being accidentally muted or sending audio to the wrong effects units, or just an unmanageable mess where you can't seem to find the one track that you can hear playing. If you want to mix effectively and efficiently, setting up the mix in an organised manner at the start is essential.

A good example of a simple mix template that can grow and evolve without hitting limitations along the way is that shown in Figure 16.1. In the mix layout, each instrument has its own track that can be panned to the left or right and can have its volume set to a suitable level. Using *busses*, similar

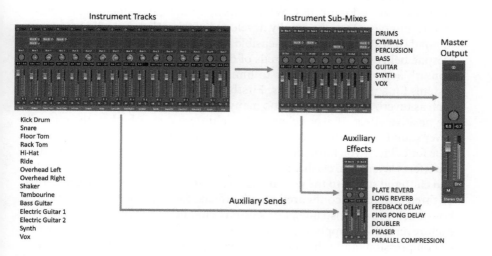

Figure 16.1 Example mix layout and structure.

instruments can be grouped into *sub-mixes*, which is essentially a new track that controls a number of instruments with a single volume control. For example, if the Kick Drum, Snare, and Tom recordings are all sub-mixed to a DRUMS sub-mix bus, we can adjust the volume of the collective instruments just by changing the volume of the DRUMS sub-mix fader. (Note that a valuable convention suggested here is to use capital letters to describe sub-mix channels, and so to differentiate them from instrument tracks in the DAW). If, however, you want to change just the sound or volume of the Kick Drum, then that can be done on the instrument track. It's really useful to use colour themes here too, giving all drum instrument tracks one colour, all guitar tracks another colour, vocals another colour, and the sub-mix channels a unique colour too; you'll be amazed how much this helps you quickly move around your mix session and find the instruments you are looking for. Using sub-mixes also allows you to easily utilise multiple instrument tracks or recording takes for a single instrument. For example, you might have recorded a first take of the drums, and then layered a second recording of tom fills afterwards. If this is the case, you will have more than one channel of Floor Tom audio. Bussing these channels to the DRUMS sub-mix essentially allows you to simplify the mix session, to the point where you may perform many of your final mixing adjustments on the sub-mix channels rather than the individual instrument channels.

Alongside the instrument tracks and sub-mix channels, we also have the ability to send audio to *auxiliary* effects. This arrangement mimics the conventional approach to mixing with an analogue desk and outboard processing units. For example, if we have a single reverb unit in the studio, we can send a little amount of each instrument channel or sub-mix channel to the reverb unit, and then mix the reverb unit's output into the master stereo output, as also shown in Figure 16.1. Whilst this approach (and much of the associated terminology) was originally developed with analogue studio equipment in mind, it still holds well as a sensible mix structure in the world of DAWs. Despite being able to use hundreds of different reverb effects in a DAW session, it's actually advantageous to limit the number of different processing systems for a number of reasons. Firstly, with the case of modulation effects such as reverb and delay/echo, too many implementations start to counteract the positive effect. If every channel has a different reverb, then the mix will never sound like anything is ever positioned in the same physical space. Likewise for delay, if you have too many delays with different timing set-ups, then the effects layer up to clash and counteract temporally, often leaving a muddy and cluttered sound to the mix. By using auxiliary sends, we can use one, two, or three differently configured reverb effects, for example, and then choose how much of each instrument or sub-mix channel we want to be effected by each reverb. If we apply this approach also to effects such as delay, parallel compression, and other more creative effects including *doublers*, *phasers*, *flangers*, and *harmonizers*, then all of the tracks in the mix get locked or glued together by funnelling through a controlled number of processing effects.

Of course, there are cases where a single instrument needs to be treated in isolation, and there are many valid instances where you may choose to apply effects to individual instrument channels or sub-mix channels.

Once a mix session is laid out with all instrument and subgroup tracks in place, it's possible to achieve an initial balance of each track's volume in the mix and for setting each track's stereo positioning. Figure 16.2 shows an example balance and panning arrangement for a drum recording session. In Figure 16.2, we can see that each drum track has its volume and stereo pan position set. For example, the volume of the hi-hat recording is set relatively low as to not overpower the mix. The overhead tracks are hard panned but also set relatively low in volume, which might be a choice made to give a closer and tighter sound to the mix, rather than one which is more weighted towards the overall sound of the kit as recorded a few feet away in the overhead microphones. The snare, toms, and cymbal channels are all panned slightly left and right from the drummer's perspective (assuming the drummer was playing a conventional right-hand kit set-up). It is just as valid to pan the microphone recordings from the audience perspective too, and different mix engineers have different preferences on this, and some will change their panning perspective of the kit for different projects.

With panning, it is often valuable to balance high- and low-frequency sounds from left to right too. It's common to keep the kick and snare drums panned close to the centre, because they are authentically positioned towards the center of the kit, but also to allow the most power and impact of these two drums which are hugely significant in the rhythm of all music genres. When panning further to the left and right, our ears tend to appreciate a balance of frequency content, so if there is lots of high-frequency content on the right, this can be balanced by some other high-frequency content on the left. This is naturally catered for with a drum kit, since usually there are cymbals

Figure 16.2 Drum instrument tracks with panning and volume adjustments, sub-mix bussing, and auxiliary sends.

positioned on both the left and the right of the set-up, but if, for example, the hi-hat is the only cymbal used throughout the song, you may find that it sounds rather unbalanced on the far left- or right-hand side, and there is justification for panning it closer to the centre. Similarly with toms, it is exciting and impactful to pan the different toms to the far left and far right of the stereo field, but in reality these extremes can sound a bit too distracting from the whole song when mixed in with other instruments, so a slightly narrower panning of the toms often works best in the final mix. This approach to balancing is also applied to the Shaker and Tambourine channels in Figure 16.2 (panned to the left and right, respectively), which may have been recorded as overdub takes after the main drum track recording was complete.

In the example shown in Figure 16.2, we can also see that the audio of the Snare, Floor Tom, and Rack Tom channels are sent to busses 8 and 9 with varying amounts of volume. In this example (though not shown in Figure 16.2), bus 8 is arranged as an auxiliary channel with a reverb effect inserted. Bus 9 is arranged similarly but with a delay effect inserted. We can see by Logic Pro's rotational send gain meters that the Snare track has a significant level of audio sent to bus 8 (reverb) and a lesser amount sent to bus 9 (delay). The Floor Tom and Rack Tom channels are sent only to the reverb effect on bus 8. Looking further at Figure 16.2, we can see the sub-mix settings for each instrument track. The outputs of Kick, Snare, Floor Tom, and Rack Tom are sent to bus 1, which is the DRUMS sub-mix. Hi-Hat, Crash, Ride, and Overheads are sent to bus 2 (CYMBALS sub-mix), and the Shaker and Tambourine are sent to the PERCUSSION sub-mix which is arranged on bus 3.

Of course, as with all aspects of drum sound, there are very few rules that must be adhered to, and a good mix engineer will experiment with many different mix concepts and approaches throughout their career. Moreover, it is valuable to have a solid intention for how you want the drums to sound before even starting the mix, and then implement the structure and organisation of the tracks to enable you to achieve the desired end result.

16.2 Dynamics processing

Drums are somewhat defined by the fact that they generate extremely dynamic and transient sound waveforms. As such, dynamics processing tools are fundamental to controlling and enhancing drum recordings in the mix. There are lots of different dynamics tools, and they all have one key feature in common; they alter the *amplitude* of the recorded signal in different ways. In many respects, the volume fader is the simplest form of dynamic processing tool, allowing us to manually adjust the amplitude levels of a track. We can either set the fader to one place or adjust it as the song goes on, to turn down any sections that have been recorded a bit too loud, or to boost sections that we want to stand out more. Of course, this is a very manual process that previously could only be conducted in real time as the song plays, but nowadays it is possible to draw volume automation commands into a DAW arrange window. Nevertheless, this still involves a very fine level of attention and adjustment. It's no surprise then that tools were developed to perform some of these volume adjustments automatically or dynamically as the song plays on. Furthermore, using automated dynamic processing tools, we can not only turn individual sounds up and down, but dynamic tools can act so fast that they are capable of reshaping the audio waveforms themselves, enabling many creative possibilities.

16.2.1 Compression and limiting

Perhaps the most used processing tool in a mix engineer's set is the *dynamic range compressor*. We will describe this in a moment, but first it's worth describing a more simple version of the dynamic range compressor: the *fast acting limiter*. A fast acting limiter looks at every single digital sample of the audio waveform and sets a limit on how large the waveform can get. In audio, we generally refer to the loudest possible digital signal level as being at 0 dB (*zero decibels*). Any signal data greater than 0 dB is technically beyond the capabilities of a digital audio system to convert into accurate sound. So, we can set a limiter's limit, or *threshold*, to be somewhere below 0 dB, say −3 dB. In this case, any audio data that is greater than −3 dB will be hard limited to a maximum level of −3 dB and no greater. As you can imagine, this can significantly alter the waveform shape of any sounds that would normally go over the −3 dB limit. Changing the waveform shape with a limiter does alter the sound of the audio, and in scientific terms adds distortion to the sound. But this distortion is not always perceivable, and so it is possible to limit a signal a little without listeners really noticing the distortion artefacts. On the flip side, it is very possible to add excessive limiting to an audio waveform; for example, if the limit were set to −12 dB, depending on the audio material, this could cause a huge amount of audible distortion to be heard on the processed audio. So why would we bother limiting if this causes distortion? Well, we've just said that sometimes the distortion introduced is relatively mild and unnoticeable, and if we have limited our waveform to a peak threshold of −3 dB, then we have the opportunity to turn up the volume of the entire audio track by 3 dB (known as *make-up gain*) without ever risking going over the maximum 0-dB level that our audio device can handle. This is great, with adding a little bit of limiting on just the highest peaks of our audio waveform; we are able to make the signal denser and louder, which can be useful in many mix scenarios.

The compressor is a more complex version of the limiter that allows much more control and creativity in the way it is used. It differs from the fast acting limiter in a number of subtle ways: Firstly, the compressor doesn't hard limit anything that crosses the threshold; instead it has a *compression ratio* setting which allows us to define how much a signal is turned down, attenuated, or compressed once it crosses the set threshold. A high ratio, for example 100:1, acts quite like a limiter, because any signal amplitude that crosses the threshold causes the audio to be turned down by 100 times (in fact, a limiter technically has a compression ratio of infinity, but any compressor with a ratio set to 100 or more can act very much like a limiter). If we use a lower compression ratio, for example 2:1, this means that waveforms that go over the threshold are turned down by two times (or halved). We therefore have some very fine control over the compressor by choosing how to set the threshold and ratio values with respect to each other. A low threshold with a shallow or low ratio will give some very gentle compression that might be quite unnoticeable, whereas a higher threshold with a steep or high ratio gives a much more drastic alteration of the waveform amplitude. Both are valuable approaches when mixing drums, and sometimes it might be valuable to have two compressors in line on a single channel with different settings applied. Sonically, the harder compression that nears that of limiting allows signals to be turned up in volume quite drastically, and sometimes the distortion created by compressing the peaks of the waveforms can add some subjective brightness and cutting impact to a drum sound. A lower threshold with a shallow ratio is very useful for adding some weight and density to a drum sound without significantly altering the sonic characteristics of the original recording. So both compression approaches can be useful for adding presence and impact to the drums in a mix in subtly different ways.

The compressor also differs from a limiter by the fact that it does not respond immediately to every digital data sample in the audio file. Instead it responds to a moving average of the signal level, which is more in keeping with how our ears and brain respond to or notice volume changes. The threshold is therefore compared against the average waveform signal, and, as such, some aspects of the waveform may cross over the threshold without ever triggering the compressor into action. We have another control here on the compressor which is to set the time window for how the average signal value is calculated.[4] If the average value is calculated over a long period of time, then the compressor is relatively slow to react (compared, for example, to the fast acting limiter) and allows momentary peaks to cross over the compressor with no effect. This is great for drums because it means that with careful setting of the compressor's *attack time*, we can decide if we want to compress every aspect of the signal (with a fast attack setting) or if we want to let a little bit of the signal (i.e. the first onset of the drum sound) through without being compressed at all (with a slower compressor attack time). The slower attack time is valuable if you want to let the natural crack of the stick hit on the drumhead through the compressor, before the compression kicks in and boosts the overall body and decay of the drum sound. The same options apply for the compressor returning below the threshold to an inactive state, so we can also set a *release time* too, to indicate how long the compression effect continues to be applied after the signal has dropped back below the threshold. Again, this is a valuable setting for drums, because it allows us to somewhat control the duration of each drum hit and its decay time, and it can have an influence on how little or how much of

Figure 16.3 Compressor plug-ins – Waves RCompressor and the SSL Native Bus Compressor.

the drum's edge overtone shines through in the decay of the waveform too. Figure 16.3 shows two software compressors, each with a number of different settings for adjustment. Every compressor has its own unique way of averaging the signal, of switching in the compression when the threshold is exceeded, and of calculating the amount of compression to apply, and each has a unique sonic signature of its own too.

There are a huge number of ways to manipulate the basic compression settings (threshold, ratio, attack, and release) when mixing drums, and you'll find lots of books and well-known engineers explaining how they use these settings for different purposes.[5,6] Some compressors also have a *knee* setting which defines a small threshold window where the compressor gradually starts to operate but not at the full compression ratio. For compressors with a knee setting, as the signal approaches the compression threshold, the ratio gradually adjusts towards its set value, rather than jumping immediately to the set compression ratio as soon as the threshold is exceeded. Given all the subtly different compressor designs, it's therefore worthwhile to experiment with many compressor types and all of the associated settings to develop your own understanding of how drum sounds can be manipulated with compression. Put a compressor on the kick drum channel and experiment with adjusting the attack and release times and the threshold and ratio values. Get an idea of how these settings affect the sound of the drum and how you might use these to enhance or control aspects of the sound in a mix. With the attack and release times, you should also be able to find a setting that allows the drums to pump or "breathe" in and out at a sympathetic rate in relation to the tempo of the song, the result of which can sound tight and impactful. But what works great or is needed in one mix might be totally inappropriate in another project. On the whole, we use compression on drums to level out volume variations in a performance, to add volume, density, weight, or presence, to soften or sharpen the attack and to shorten or lengthen the decay, and often to add some subtle distortion or sonic character to a drum recording too.

Grammy award-winning mix engineer Simon Gogerly (U2, No Doubt, Underworld) explains his approach to using attack and release times, referring to affecting the *transient* or attack of the drum signal, and also the *body* or sustain and decay of the drum sound too:

> I might have some overall compression on overheads with fast attack and slow release, because I don't want the transients to come through the cymbals too much. But with kick and snare I'll have slower attacks, so that the transient comes through, and I'll adjust the release depending on the kind of body I want in the sound, and also based on the tempo of the track. Especially with the kick drum, because I want the kick drum compressor to open back up in time for the next beat, which helps the kick sound more rhythmic in the track.[7]

So far, we've talked about adding compression to a single track, i.e. to the kick, snare, or toms. But we can add compression to the sub-mix busses too. We might do this if we want to apply one dynamic control equally to a number of instruments in one simple process. Or we might add compression to the drum sub-mix as part of a staged compression chain, for example, so the kick drum is compressed once on its own channel (perhaps with a low threshold and low ratio) and then compressed again as part of the drum sub-mix (perhaps this time with a relatively high threshold and a high ratio). It can also be valuable to compress the cymbal recordings too, either individually or more often on a sub-mix of the cymbals. Adding some subtle compression to the cymbals brings a crispness and brightness to their sound and allows their quieter decay tails to be brought up in volume to a more noticeable and charismatic level. Some mix engineers like to hard compress room mic recordings too, and subtly blend in a hyper-compressed element that gives some enhanced character of the recording space that was used.

It is also possible (and recommended) to add compression in a parallel form. Parallel compression refers to the case where the compressor is positioned on its own auxiliary effects bus and is then driven through the DAW send busses, in an identical way to how the reverb and delay effects were set up in Figure 16.1. Parallel compression allows some quite drastic compression to be applied, but to be blended in more subtly alongside the direct tracks and sub-mixes. The result is that parallel compression applied to drums allows the overall volume and density of the drums to be increased from the ground up, without significantly increasing the peak level of the drum tracks. With parallel compression on drums, it's sometimes advantageous to have one parallel compressor for the kick and one for the snare, and another for all of the other drums or instruments in the mix. When feeding a snare or a kick track into a parallel compressor, these very dynamic sounds impact the compressor settings as they are heard, so they can often cause an unwanted pumping

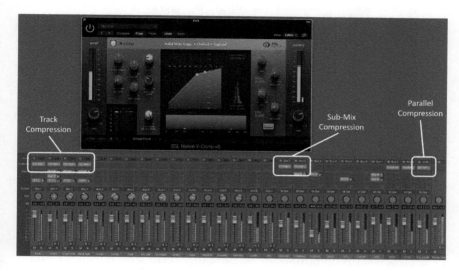

Figure 16.4 Track, sub-mix, and parallel compression used on drums in a mix.

effect if using a single parallel compressor for multiple instruments (of course, this might be wanted also for some styles of music!). If the kick and snare have their own parallel compression channels, then this can be very effective for blending strength and density into the mix without the risk of causing distortion. If you are mixing metal music, then it's not uncommon to send a parallel compression channel to a second parallel auxiliary with compression applied too, to achieve a two- or three-tier parallel compression process that adds more and more impact to the drums with each extra stage. Of course, it's possible to go too far and result in everything sounding like a muddy distorted mess, so a creative judgement needs to be applied with each decision to add extra compression. Figure 16.4 shows the in-line compressor inserted on the Kick Drum track, though it is also possible to see further compression effects are applied to both the DRUMS and CYMBALS sub-mixes (on busses 1 and 2, respectively) and parallel compression set up on bus 10.

Mix engineer Simon Gogerly emphasizes his personalised approach to parallel drum compression, incorporating a hardware valve distortion unit:

> I usually setup a parallel bus compressor with a fairly fast attack and fast release. My parallel setup is perhaps a little different because I use a valve distortion unit after the compressor – the Thermionic Culture Vulture – which adds additional compression and harmonic distortion. The great thing about the Culture Vulture is that it allows either even or odd harmonic distortion, and one or the other will suit different

types of music. I then mix the "dirty" parallel channel back in with the "clean" mix of drums. This allows transients and the attack of the kit coming through from the track compression going to the clean mix bus, and the body of the drums and extra harmonics coming through from the parallel channel. To get those to blend just right, I usually add some overall bus compression on the clean drum mix too. [7]

TRY FOR YOURSELF: TRACK, GROUP, AND PARALLEL COMPRESSION

Set up a mix of a multichannel drum recording similar to that shown in Figure 16.4, implementing track, sub-mix, and parallel compression processing. First, use the track compression on the kick, snare, and toms to add an overall weight and density to the drums; for this you may want to set a fairly low threshold (−20 or −30 dB) with a moderate compression ratio (around 1.5–2.5). Experiment with the attack and release thresholds to keep the drums sounding natural and crisp, but with improved power.

Now experiment with adding sub-mix compression. For the drums, you may want a fairly high threshold (e.g. −10 dB) with a harder compression ratio (e.g. 3–5). Of course, the values used will depend on the signal levels of your recordings. A lower threshold with softer ratio may work better for the cymbals, just to add a little additional brightness to the mix.

Now experiment with the parallel compression, and set this quite hard, with a relatively low threshold and a fairly high ratio. This will cause some distortion artefacts in the parallel sound, but you can use the send gain or auxiliary volume fader to adjust the amount of volume of the parallel compression channel in the overall mix.

Experiment with all these types of compression and look for an overall setting that takes your initial drum recordings to a new level of power, crispness, and impact. You may even want to add a second parallel compressor and send the snare and kick to different parallel channels.

16.2.2 Gates and expanders

Another extremely useful dynamics tool for mixing drums is the *dynamic gate*. A gate is quite a simple sound processor that works similar but in an opposite method to the limiter. With a gate, when there is only a very small amount of signal, the track output remains silent. So the signal needs to exceed a certain volume or threshold in order to open the gate and be heard. This is a valuable tool for automatically removing unwanted microphone bleed or background sounds from a drum track, for example, to keep a tom

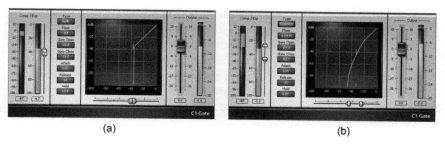

Figure 16.5 Waves C1 plug-in in (a) gate and (b) expander modes.

recording silent except for only the times when the tom is played. Despite its simplicity, setting a gate on drum tracks can be quite challenging in practice and requires a good attention to detail. This is because, for example, it is very easy to set a gate to work on a few chosen snare hits, but to not realise that it has inadvertently also muted all the ghost drumstick notes and other performance nuances on the recording. Additionally, the gate can be a very binary tool – there is either loud sound or no sound and nothing in between. Whilst this is the main purpose of the gate, and the audible pumping effect suits some styles of music well, it can often be more beneficial to use an *expander* effect for the same purpose. An expander is to a gate what a compressor is to a limiter, it essentially is a gate without such binary on/ off settings and a more graded transition between the fully on and fully off states. An expander can therefore be set to attenuate or reduce the volume of sounds that drop below a threshold, and, as with a compressor, it is possible to set a ratio for how strong the attenuation is applied. We also have attack and release settings with an expander which allow the responsiveness and timing of the attenuation to be controlled and optimised. Figure 16.5 shows the Waves C1 Gate which can also be set to work in an expander mode.

TRY FOR YOURSELF: GATES AND EXPANDERS

Take a recording of a kick drum microphone and apply a gate plug-in to the channel. Experiment with the gate settings to allow just the sound of the kick to be heard and to give silence at all other times. This might be quite easy if the recording is very well isolated, but you might find it a challenge to isolate just the kick without also isolating the snare or other drums that were being played at the same time. Try adjusting the attack and release times of the gate to achieve a better isolation of the kick drum sound. Listen to the end result in the context of a full drum mix; in some cases, the gate can clean up the sound

(Continued)

of the kick to make it quite prominent and clear, whereas in other circumstances, the kick may start to sound like it is pumping unnaturally and switching on and off in a harsh and overly drastic way.

Now experiment with an expander plug-in in the place of the gate and see if you can achieve a more satisfying result. The kick will not be as clean or isolated as with a gate plug-in, but may give a more natural on and off feel when it is blended back with the other microphone channels.

Try the same approach with the snare drum. Sometimes the benefits of the gate or expander are valuable, but sometimes the bleed from other drums in each microphone is acceptable (or even beneficial) and the drum tracks may not need gate or expander processing at all.

16.2.3 Envelope shaping

It's a valuable skill to master all aspects of compression, limiting, gating, and expansion for mixing drums. Drums are dynamic beasts and they can often only be tamed, and their full power harnessed into a useable form, with dynamic processing tools. On the positive side, with dynamics processing, it is possible to make drums sound even bigger, bolder, stronger, more impactful, more exciting, and more in context with the song, or hyper-real, i.e. even better than in real life! Whilst there are some very subtle settings and characteristics to master, there are also a number of tools that simplify the approach to dynamics control, by implementing their own algorithms that make drums and other instruments sound fatter, crisper, and more impactful. One example, used by many mix engineers, is the SPL Transient Designer which particularly focuses on enabling sounds to have a sharper or softer attack or a longer or shorter sustain and decay sound, as shown in Figure 16.6. These dynamic elements of the signal are often collectively referred to as the *envelope* of the waveform. A further example of an envelope shaping tool is the Waves Smack Attack plug-in, also shown in Figure 16.6.

16.3 Hybrid mixing

We've only discussed software tools for mixing and processing recorded sounds, but it's important to know that these often represent software recreations of physical equipment that was invented long before computers and digital audio workstations existed. For example, analogue audio compressor units were developed in the early 1900s for increasing the power of signals that were being broadcast over the analogue radio spectrum. After the Second World War, some of these units were adapted by audio pioneers to

 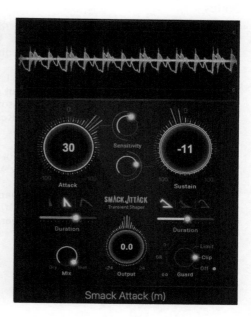

Figure 16.6 SPL Transient Designer and Waves Smack Attack effects for envelope shaping.

be used in music production, enabling the dynamic range compression effect that is so creative and useful on drums, vocals, and all music. The original electronic hardware relied on valve signal amplifiers and opto-resistor circuits for detecting the audio threshold, which inherently created a unique sound that has contributed to the modern aesthetic of popular music. It's still very possible to acquire new and original hardware compressors – such as the Teletronix LA-2A or the Universal Audio 1176 – and many engineers insist on the sonic benefits of such analogue processing hardware over similar software emulations. Nowadays, the modern DAW and mixing consoles allow a best-of-both-worlds hybrid approach, enabling outboard compressors, equalisers, distortion units, reverbs, and summing channels to be integrated with a software platform and digital tools, which themselves hold unique benefits and capabilities. The subject of hybrid mixing, incorporating analogue audio equipment into a digital audio workstation, is worthy of a book of its own, with so many different approaches, options, and set-up topologies. Here, therefore, we will continue to discuss audio processing exclusively from a software perspective. If you get chance, however, to experiment with a valve compressor, a distortion unit or a sought-after audio equaliser, then it is well worth taking the opportunity to experience exactly what values and benefits can be achieved from implementing hybrid mixing with drums.

Notes

1 *Mixing Audio: Concepts, Practices, and Tools*, 3rd Edition, Routledge, 2017, by Roey Izhaki.
2 *The Mixing Engineer's Handbook*, 4th edition, Bobby Owsinski Media Group, 2017, by Bobby Owsinski.
3 *Mixing and Mastering in the Box*, Oxford University Press, 2014, by Steve Savage.
4 There are actually a number of different designs and methods for implementing attack and release times in compressors; some are evaluated and explained in more detail in F. Floru, Attack and Release Time Constants in RMS-Based Feedback Compressors, *Journal of the Audio Engineering Society*, vol. 47, pp. 788–804, October 1999.
5 For example, some detailed examples of how compressor attack and release times affect the audio signal are given by Alex Case in his book *Sound FX*, Focal Press, 2007, pp. 153–8.
6 For example, Mark Mynett discusses many aspects of mixing drums for metal and hard rock music in his book *Metal Music Manual*, Focal Press, 2017.
7 Interview with mix engineer Simon Gogerly conducted on 06/10/2020.

17 Mixing drums
Creative processing

In Chapter 16, we discussed setting up a mix and considering different approaches and tools for manipulating volume, panning, and the dynamic profile or envelope of drum sounds in a mix. Having crafted a mix that sounds balanced and clear, it is possible to use creative processing tools to develop further impact, excitement, clarity, and spaciousness. There are many bespoke, weird, and wonderful tools available, though most can be broken down to being from one of the following categories:

- Spectral processing (including filters and equalisation)
- Spatial processing (including reverberation and delay)
- Enhancers and exciters (including distortion and harmonic enhancement)

Of course, it's possible to combine approaches within these categories, which many audio plug-in developers do to realise novel, creative, and more abstract audio processing tools. It's also possible to combine these tools yourself in parallel or series or with *sidechain processing*, to open up a huge number of approaches and techniques for creatively manipulating the sound of recorded drums.

17.1 Equalisation and spectral processing

Alongside dynamics processing tools as the most widely used mix effects are spectral processing tools. The most common spectral processor is the *equaliser* (*EQ*), which allows us to manipulate the volume of different aspects of a waveform's frequency spectrum. The equaliser is rarely used in mixing for making things "equal", though. That was its original intended purpose, since some early audio systems (record players, amplifiers, and loudspeakers) were incapable of playing out all frequencies equally, so the audio equaliser, which is essentially a network of variable electronic audio filters, allowed some manipulation of the sound to counteract these adverse effects. In mixing, we tend to use an equaliser to make things less equal nowadays and to give different recording tracks specific characteristics that allow them to shine and stand out, and avoid clashing, competing, or masking with respect to other recording tracks.

The standard *parametric equaliser* allows a number of frequency bands to be adjusted and has settings of frequency, bandwidth, and gain. The frequency setting is self-explanatory, allowing us to choose which frequency in the range of human hearing (20–20,000 Hz) is to be manipulated. The bandwidth setting is usually denoted Q, since it is the term of a mathematical equation for widening or tightening the frequency band which is being manipulated. Gain can be set positive or negative to either boost or attenuate the chosen frequency band. For frequency bands that are at the top or bottom end of the range, we usually have a choice of EQ shape, because the lowest and highest bands do not connect with an adjacent band outside the audible range. For the edge bands, we can use either a *full-cut* or a more gentle *shelf*-shaped EQ. The full-cut setting uses what are called *high-pass filters* (allowing high frequencies to pass and low frequencies to be cut) and *low-pass filters* (allowing low frequencies to pass and high frequencies to be cut). The shelf filter can be set similar to a cut filter, except it does not attenuate the rejected frequencies completely, instead allowing a gain to be applied to decide how much the filter band is reduced or amplified by. Figure 17.1 shows a six-band equaliser that has a number of different EQ filters implemented on a snare drum channel. First, we see that band 1 has a low-cut (or high-pass) filter applied at 100 Hz. The gain setting has no effect on a cut filter, but the Q value allows the steepness of the filter to be altered. Note that it is not possible to set a cut filter (or any audio filter) to have an infinitely steep or *brickwall* profile – this is because the practical electronic and mathematical theory for filter design does not allow it. It is possible to set the Q value to be quite steep (or narrow for parametric bands), but be aware that steeper

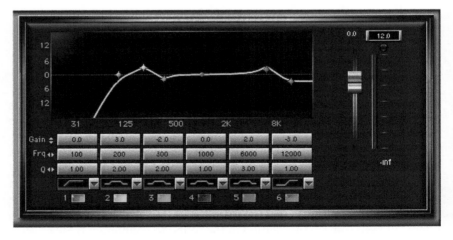

Figure 17.1 Six-band parametric equaliser applied to a snare drum recording (Waves Renaissance EQ plug-in).

Q values push the mathematical filter theory more to its limits and the most steep filters can often cause other unwanted sonic artefacts to be generated as a result (such as ringing, phase incoherence, or peaks and dips at other frequencies). Looking further at Figure 17.1, we see that bands 2 and 5 are set as parametric boosts, whereas band 3 is set as a parametric attenuation. Band 4 is not switched on in this example, and band 6 is set as a high-shelf filter with an attenuation of 3 dB (or, equivalently, a gain value of -3 dB).

It's useful to have an understanding of when and why we might use these frequency bands in order to fully explain the example given in Figure 17.1. There are many reasons why we might use EQ on drum tracks and sub-mixes; here are a few which are discussed below.

17.1.1 Cutting low frequencies

Cutting low frequencies might seem counter-intuitive for an instrument that occupies the low-frequency range, but usually drum recordings that have used a number of different microphones and microphone positions will benefit in clarity and presence if the low frequencies towards the bottom range of human hearing are carefully controlled. With all the different microphones, drums which sound great in individual mics can combine to cause a muddy, dull mess when all are played back at the same time. Cutting low frequencies can have a very positive effect on this, predominantly by reducing the amount of phase-related clashing when the microphone channels are all mixed together and reducing the build-up of low-frequency room reverberation captured in each mic and compounded when the channels are mixed together. It's extremely valuable therefore to apply a low-cut filter to most or all drum recordings and choose carefully where to set the low-cut frequency, which should generally be lower than the fundamental frequency of the drum being considered. We know that kick drums have a fundamental frequency of around 50 or 60 Hz, so it makes sense to remove frequencies below about 50 Hz, since this content can only be low-level rumble and low-level reverberation in the room. With the snare, we have the same situation, but also with low-frequency bleed from the kick drum into the snare mic too, so if our snare is tuned to 200 Hz, we know we can safely set a low-cut EQ on the snare channel at, say 100 Hz (as set on EQ band 1 in Figure 17.1). This low-cut filter is also removing low-frequency room sound and attenuating the bleed of the kick in snare mic too, reducing the chance of phase coherency issues when we blend the two microphone sources together in the sub-mix. Using low-cut filters on the cymbal, overhead, and room mics is often beneficial too, either directly on each microphone channel or on the sub-mix channels. If you are looking for a very crisp sound of each instrument in the kit having its own clear definition in the mix, then low-cut filtering the cymbals will work well. However, sometimes the overhead or room mics capture some valuable sonic qualities of (particularly) the snare drum, which are not evident in the snare's spot mic recording. So you may want to set a low-cut filter on your cymbals' sub-mix and gradually increase the frequency, listening careful to the full mix to hear how the snare sound changes as you take the filter higher. If you prefer the sound of the snare's close mic, then you can take the cymbal cut all the way up towards 500 Hz, but if you want more of the snare's natural room sound in the mix, then you'll need to keep this lower at around the snare's fundamental frequency of 150–200 Hz.

17.1.2 Treating the fundamental and overtones of each drum

If you know the fundamental tuned pitch of each drum in your kit, then it is very easy to set a parametric EQ on this frequency and boost the fundamental to give more emphasis and a slight exaggeration to each drum in the mix. This works better than just turning up the whole track sometimes and gives a more focused boost to the particular drum that you want to stand out. As can be seen in Figure 17.1 (EQ band 2), the snare drum's fundamental frequency of 200 Hz has been boosted by 3 dB. This approach can be applied similarly for the drum's edge overtone, which we know will be about 1.5–1.7 times greater than the fundamental. Often drummers and studio engineers prefer to reduce the snare drum's overtones both during recording and in the mix, because they can sometimes ring out longer and distract from the more powerful and lower fundamental pitch of the drum. We can manipulate this with EQ therefore and reduce the overtone frequency a little, as, for example, in Figure 17.1 (EQ band 3), which shows 2 dB of attenuation at the snare drum's edge overtone frequency of 300 Hz.

17.1.3 Adding attack and presence

It is possible to boost or cut the more timbral frequencies of the drum too – these are those frequencies associated with the characteristics of the drum's design, construction, and materials. We know that drums vibrate subtly at many different frequencies all the way up the frequency spectrum, with vibration of the drum shell, the metal hardware, snare wires, and the mounts all adding that special magic to the sound of the drum. Whilst it's hard to know what these exact frequencies are, we can make some broad brush strokes with the EQ to make the drum feel brighter or warmer, or to bring out some character or presence in the sound. The metallic hoops and hardware ring out into and above the 1,000-Hz range, and we also know that high-frequency content relates to the attack and sharpness of a musical sound too, so if we want to make a drum sound brighter and cut through the mix, then we might want to add some EQ boost at a frequency between 2 and 8 kHz. It's never quite obvious which frequency will give the boost you are looking for, so set a parametric EQ boost on the snare, for example, and sweep it up and down the frequency range until you hear a point where an impactful characteristic of the snare is being enhanced. It's very easy to overdo this and boost too much, making the drum sound thinner and unnatural, but with careful adjustment you can add some valuable sonic presence to the sound when considered in the full mix. This setting is seen in Figure 17.1 (EQ band 5), which shows a presence boost of 2 dB at 6 kHz. This effect works also for kick and tom drums, though to a lesser extent because of the reduced amount of high-frequency content from kick and tom drums in comparison with a snare. Equally, be careful to check you have found the right solution for the change you are looking to make; it might be that a compressor or envelope shaping tool gives a better result for enhancing a drum's attack character rather than a high-frequency EQ boost.

17.1.4 Controlling high frequencies

As with controlling low frequencies, it is also often valuable to control the high frequencies in multi-microphone recordings of drums. There is usually less issue with phase-related artefacts at the high end of the frequency spectrum, but cymbals themselves cause significant bleed in drum recordings resulting in two challenges. Firstly, it is hard to manipulate the relative balance of sounds in the mix; this is common when there is a lot of hi-hat sound captured in the snare mic, for example, and it becomes impossible to enhance the snare sound without causing the hi-hat to become too loud in the mix. Secondly, if cymbals are captured on all of the microphone tracks, then it becomes much harder to create a wide stereo image of the drums in the mix. Cymbal sounds captured on the kick, snare, and tom channels cause a significant amount of cymbal sound to become evident in the centre of the mix, and this diminishes the stereo effect gained by hard panning overheads and close cymbal mics to wide parts of the stereo field. Subsequently, it can be valuable to reduce the high frequencies in drum tracks that do not need so much high-frequency content. Figure 17.1 shows a high-frequency shelf (EQ band 6) at 12 kHz. This is applied, in this example, to reduce the impact of the hi-hat on the snare channel, allowing the snare to be compressed, boosted, and enhanced without causing an increased presence of the hi-hat in the mix. There is clearly a trade-off here, as in many mix tasks, since the snare would undoubtedly sound a little more natural and clear without this EQ shelf applied. But without the high-shelf filter, the hi-hat would become enhanced too much by the compression applied to the close snare mic. If a spot mic recording of the hi-hat is available, then this can help reduce the impact of applying a high cut or high shelf to the snare.

Grammy award-winning mix engineer Simon Gogerly gives some insight into how he uses EQ with drums in order to achieve a more modern and contemporary sound, whilst also emphasizing some EQ pitfalls to avoid:

> Bass drums on old records don't have a lot of sub frequencies and snares don't have much high end and crack. But modern tools allow drums

with a wider, "hyper-real", frequency range to be crafted in the mix. I try to fit each drum into its own frequency band, from low to high; kick, toms, snare, hi-hats and cymbals all have their own fundamental pitch and frequency range. The frequency ranges all overlap, but I use EQ to help them sit together with their own identity too. For example, I'll usually put high-pass filters on the overheads and room mics, to avoid any boominess from the kick coming through and to reduce the likelihood of low-end phase issues. You do have to be careful when using EQ with drums, because most plugins change the phase relationship of the audio and it's possible to lose some frequencies if you're not listening to the full effect of each process you add.[1]

There are clearly many scenarios and possibilities for using equalisation when mixing drums and other instruments. In many respects, the equaliser is used to boost characteristics of different instruments to make them stand out in the mix, whilst also being used to reduce some of the unavoidable compromises made through recording with multiple microphone setups, which generate bleed and phase-related issues when summed together. There are many types of equaliser too, based on both original electronic circuit designs and more modern mathematical algorithms too, and all can be used for both clinical and creative types of sound manipulation. A good approach is to avoid using excessive EQ where possible, firstly to avoid significantly noticeable and unwanted artefacts generated from extreme gains, cuts, and Q values, and secondly as a challenge to improve your recording skills and better achieve the sound you are looking for before the mixing session starts. Additionally, it is quite easy to add EQ to a single audio track and improve the sound, but the greatest benefit of EQ is predominantly on the effect you can make to a sound when heard within the complete mix; so, whilst it is useful to solo tracks and evaluate them in isolation, it is usually most valuable to apply EQ while listing to a full mix or a sub-mix of instruments all at once. Equalisation is hence an extremely powerful tool for shaping and crafting the mix, but with such great power comes great responsibility!

17.2 Using reverb to regain authenticity

Reverb is used significantly in mixing and has the ability to bring extra life and excitement to drums. However, where we may generally apply reverb quite liberally to vocals, strings, piano, guitar, and brass, the use of reverb with drums can have just as much a negative effect as positive on the mix, if not applied carefully.

Reverb allows us to bring a sense of space and location to the recording. It is often the case that close or spot microphone recordings do not capture an accurate sense of reality, since every sound we hear in the world we experience is affected by some form of reverberation and echo. But with a recording using close microphones, the balance of reverb and direct instrument sound is heavily weighted, unnaturally, towards the instrument. You have most likely never listened to a snare drum from 5 cm away or a kick drum from the inside – this would damage our hearing and sound very different to the natural sound we are used to when hearing a drum kit played from a few feet away. This principle applies equally to recording vocals, strings, guitar cabinets, and most instruments that are recorded with microphones much closer than we would listen in a real setting. The benefit of clarity and focus from a close-microphone recording is at the expense of capturing a natural sound that we are accustomed to from the real-world scenario. As a result, we usually need to add artificial reverb to these recordings to bring back a sense of reality. This is not a bad thing: we have much more control over an artificial reverb sound, and we can even use this to make things sound bigger and better than reality or hyper-real as we mentioned earlier.

Actually, drums are one instrument that we do record from a fairly natural distance, particularly overhead and room microphones placed further away from the kit, so we have already captured a natural ambience of the room in our recording, and it may be the case that no further reverb is needed on the drums in the mix at all. That said, it is very possible that the room used for recording was compromised, with poor acoustic design, heavily damped, or with random reflective surfaces or close walls and a low ceiling – giving a natural reverb that just doesn't sound as good as we hoped. In this case, we might create a mix more balanced towards the close microphone recordings with the overheads and room mics mixed quite low, so adding artificial reverb is required to bring back some reality to the drum sound. For some recordings, it may be that the room was very dry with high levels of sound absorption that left no sense or reverb or reflection in the recordings. In this instance, we may also need to add reverb because, even with the room and overhead microphones quite loud, there still is no real sense of space or location in the mix. Another reason for using artificial reverb with drums is to allow the whole song to blend together; if all of the instruments in the mix are sent to the same reverb effect, then it starts to sound more like the instruments are located in the same performance space, and they share the same sonic attributes, bringing a more natural sound and a more cohesive feeling to the full mix.

There are a number of different reverb type effects, and nowadays it's very rare for a studio to use a physical reverb chamber or a big reverb plate, so predominantly plug-ins or digital hardware reverb units have become the norm. Many plug-ins still use the concepts of physical reverb spaces or systems, and many are designed to be accurate emulations of real spaces too. Figure 17.2 shows two different reverb effects: one modelled on the sound of a physical reverb plate and the other modelled on the sound of a concert hall space.

Figure 17.2 Lexicon PCM Plate and Waves Renaissance Reverberator reverb plug-ins.

Two key measures for reverb are the *reverberation time* and the *pre-delay*. The reverberation time defines how long the reverberation goes on for and gives a sense of how reflective the room is. The pre-delay gives a sense of how big the room is, by delaying the reverb from starting for a short duration, mimicking the time that it would take for sound to travel to the walls and back. By setting these two values, we can give a sense of a real space to drums. Because drums are so transient and time based, it usually doesn't sound so good to use reverbs with pre-delays; it generally works well for drums to have the reverb tightly connected to the impact of the drum, and often adding extra pre-delay can cause muddiness to the sound and can interfere with the timing and the feel of the drum beat being played. Reverb time also can have a relationship with the drum beat – fast tempo songs generally need a shorter reverb time on the drum sounds, to avoid the reverberation of one snare hit blending into the next hit. In many contemporary music styles, we like the sound of drums to be quite tight, so longer reverb times often don't work so well with drums. For a similar reason, *early reflections*, which are the very first sound reflections to come back to us after a sound is made, can also interfere with the clarity of the recording. So if a reverb effect has an "early reflections" setting, it can sometimes be beneficial to reduce this setting (as shown in Figure 17.2) for drums. This is one reason why the reverb plate type works well for drums, particularly on a snare drum, because it is not actually an emulation of a room-type characteristic, and instead looks to emulate the reflections of sound reverberating within a solid plate of metal.

Reverb can work well on toms, even with fairly long reverb times to give a big sense or space on tom fills (as is synonymous with many classic rock recordings), but reverb on the kick drum rarely gives an improvement to the sound, unless a very specific artistic effect is desired. We often spend a long time trying to get the kick to sound strong, powerful, and tight, so reverberation often counteracts this initiative. Long reverb times can also work well on cymbal recordings, as long as the low-frequency drums have been filtered away and are not being effected by the reverb that is applied.

Reverb on the cymbals can add a valuable shine to the kit and bring a wash of high-frequency sound that the drums sit within, which compliments some musical styles. In all cases, it may be valuable to add EQ to the reverb auxiliary channel, to remove the very lowest- and very highest-frequency reverberation sounds. In many respects, reverb is almost not meant to be heard; it is something you perhaps should not notice, but becomes very noticeable when it is omitted or missing. Hence, adding EQ to the reverb channel, with high- and low-cut filters (e.g. low-cut filters up to 200–500 Hz and high-cut filters down to 6–10 kHz) can allow more reverb to be added without it becoming blatantly obvious in the mix, enabling the feeling of a natural and interesting performance space, without sounding like a significant artificial effect has been added.

17.3 Delay for drums

Delay is another tool that is used substantially when mixing vocals, guitar, synths, and much electronic music, but requires very careful consideration when applied to drums. A delay effect generates an echo of a sound at some point in time after the original sound. If applying delay to drum sounds, then it is usually critical that the delay time setting is in time with the tempo of the song, to avoid rhythmical clashing and a loss of control of the song's beat. One specific use of delay is as a timed *slapback*, which refers to an echo of a sound at a time that is in keeping with the song tempo. For example, we might set a slapback echo on the snare at an eighth-note duration after the original sound. The echo sound is set to be quieter and sound like it is "slapping back" off the back wall of a concert hall perfectly in time with the performance. Often the echo is turned down in volume so much that the slapback becomes barely audible, but gives a stronger sense of rhythm and timing to the song in a fairly imperceptible way. Additionally, the slapback sound can be treated or equalised to have a slightly brighter or duller sound depending on whether it is intended to stand out or blend in more within the song. The slapback can be just a single echo or can repeat twice or more, and many echo or delay effects have a feedback setting, which allows the echo to continue indefinitely at the set delay time, but at a gradually reducing volume until it is no longer audible. It's also possible to set the delay sounds to move in the stereo filed from left to right, so perhaps the first echo is heard on the left and the second echo is heard on the right, and they alternate thereafter – we call this type of stereo delay a *ping-pong* delay, because it mimics the movement from left to right similar to that of a ping-pong ball in a table tennis game.

Mix engineer Simon Gogerly gives an example of when and why he uses slapback delays with kick and snare drums:

> I often like to use a fairly subliminal slapback delay on snare and sometimes kick as well. This works well if the drum pattern is sparse, because

the rhythm of the track can be enhanced with a little repeat on the snare or kick. The slapback repeat usually needs to be heavily equalised so that it's not taking up too much room, so it becomes just perceptible as something that's helping to drive the rhythm a little stronger.[1]

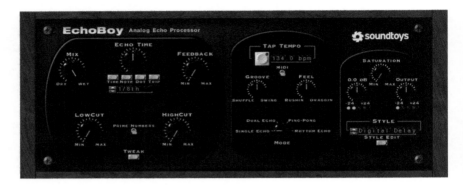

Figure 17.3 Soundtoys EchoBoy delay effect.

Figure 17.3 shows the Soundtoys EchoBoy delay effect which enables the song tempo to be set, as well as the selection of the echo time in either milliseconds or as a note length of crotchets, quavers, semiquavers, dotted beats, or triplets. The EchoBoy effect also shows a setting to cut the low- and high-frequency sound of the delay signal, and to set as a single slapback, dual echo, ping-pong, or feedback-level controlled rhythm echo too.

TRY FOR YOURSELF: SNARE SLAPBACK, FEEDBACK, AND PING-PONG DELAY

Take a snare drum track and add an auxiliary slapback delay channel, routing some of the snare sound to the auxiliary. Set the delay so that it plays back an echo of the snare hit after exactly an eighth-note of the song tempo (you might prefer to set this to a quarter or sixteenth depending on the song tempo and the beat being played). Depending on which plug-in you are using, you may need to specify the song tempo or set the delay time manually; in either case, set the slapback echo so that after each snare hit you hear a second snare hit in time with the song. Now experiment with volumes; can you set the delay volume so that it is barely noticeable in the mix but gives a subliminal rhythmic enhancement to the track? Now add a feedback setting to the delay and hear how each snare hit creates a number of consecutive in-time echos as it decays to silence.

Expand the set-up to incorporate a stereo ping-pong effect; you should be able to do this in a single stereo delay plug-in. Can you set the delay to give two echos of every snare sound? Space the echos so that the first one occurs on the left and the second one occurs on the right (and the initial hit in the centre). You can also experiment with the timing, so that the echos are in time but not conventionally on eighth notes. For example, one side of the ping-pong might be timed to occur after three sixteenths, which will give an interesting rhythmic characteristic. This is a good exercise to show you have control over your delay plug-ins, even if you don't want a ping-pong effect on your snare drum!

17.4 Distortion and enhancers

In aiming to make drums sound bolder, bigger, and better than reality, mix engineers have many tricks and tools up their sleeve. There is no single approach to mixing drums; the skill of a mix engineer is in assessing the quality of the recordings, evaluating these against the aim of the mix, or the genre of the song, and deciding what effects will achieve the desired result. If the recordings sound dull or don't seem to cut through and have presence against the other instruments in the mix, then some subtle distortion or harmonic enhancement might be the solution you are looking for. We all know the sound of creative distortion, applied to an electric guitar to give a classic rock, blues, or metal feel, bringing warmth, edge, or drive to the instrument sound. We can achieve a similar result with drums by applying a distortion effect too, though somewhat more subtly. Because drums are so dynamic and only have a high signal level for a short period of time, it's possible to add distortion to drums without the effect being adversely noticeable within a mix. A little distortion added to a kick drum or a snare adds a momentary brightness and edge to the sound, and while this might sound a little unnatural or "crunchy" when heard in isolation, within the mix it can give a more subtle characteristic that allows the drums to stand out more without needing to be significantly louder. There are many distortion plug-ins available, mostly designed for use on guitar, though some work better than others when applied to drums. Alongside the much-loved Empirical Labs Distressor, the Soundtoys Decapitator (as shown in Figure 17.4) is one such distortion effect that works well on drums, allowing enough settings to find a modified sound that enhances and stands out without causing an unpleasant artefact. The drive setting allows a subtle amount of distortion to be chosen, and the EQ and tone controls allow this to be blended to suit and compliment the sound of the drum being processed. A number of different distortion styles can be selected (labelled A, E, N, T, P, and Punish), and the mix setting allows the blend between the amount of affected and unaffected

Figure 17.4 Soundtoys Decapitator plug-in.

signals to be set. Where more simple distortion plug-ins work well on most guitar sounds, often this more detailed level of control is required to achieve a suitable distortion sound when applied to drums. A little distortion on drums can work well with all popular music genres, from pop, hip-hop, and R&B, to indie, rock, and metal.

A close relative to subtle distortion is the harmonic enhancer effect. The designs of these are often steeped in mystery, as each enhancer performs its own special magic on an audio signal. Distortion itself causes harmonics of the signal to be created, which can be musically pleasing if applied subtly. Harmonic enhancers work on the principle that some harmonics generated by distortion sound more musical than others (it's generally regarded that even harmonics sound more musical than odd harmonics for example), so harmonic enhancers build on this concept with their own unique formula. For this reason, you'll rarely find many settings on an enhancer plug-in, and often just a simple dial to set how much enhancement is applied. In a similar way, it's also possible to add low-frequency sub-harmonics and resonant bass frequencies related to the audio through electronics or digital algorithms too. Sylvia Massy explains a valuable use for harmonizers, when mixing drums:

> For some types of music I will add a harmonizer across the overhead mics, tuning them down a full octave. This will deepen the overall sound of a kit.[2]

So we can use exciters and harmonizers that work above the fundamental frequency of the source audio, to add shine and sparkle, and those which work at lower frequencies than the source audio to add bass, fatness, and sub-resonances to the sound. Two such examples of audio enhancers

Figure 17.5 Aural Exciter and Little Labs Voice of God effects.

that work well on drums are the Aphex Aural Exciter (harmonic exciter) and the Little Labs Voice of God (sub-bass exciter), both shown in
Figure 17.5.

17.5 Sequential and sidechain processing

Sequential and sidechain processing refers to different methods of combining
audio effects into a single hybrid effect. This can be as simple as adding an
equaliser in series with a reverb or delay effect. In fact, it's usually valuable
to add equalisation to all reverb and delay effects, because as they are used
more and more in your mix, they either become very obvious and stand out,
with some high-frequency harshness, or cause low-frequency build-ups that
make the mix sound muddy. Adding an EQ to the reverb or delay channel allows you to shape the sound and make your effects more subliminal and impactful. You can be creative too; for example, try adding a phasor or flanger
to a reverb channel for some interesting subliminal movement in the mix.
The classic gated reverb sound – discovered by accident by Peter Gabriel
and Phill Collins of Genesis – follows this approach too. The gated reverb
is exactly what it says, a reverb with a gate, the result being an obvious and
impactful reverb sound that is applied to drums, but is gated to cut to silence
after a very short time. You can hear this sound on most classic rock drum
sounds from the 1980s, and it's still used regularly today, though perhaps
more subtly. When set up well, the effect can add a big presence and power
to snare and tom sounds without a long reverberation tail that muddies up
the rest of the mix.

TRY FOR YOURSELF: GATED REVERB

Create an auxiliary reverb channel with a big hall or church reverb setting. Send some of the snare or tom sounds and hear the big reverb in action. Now add a gate plug-in directly after the reverb plug-in on the auxiliary channel. Set the gate threshold to cause the reverb to snap to silence after a short burst of sound. You might get a better result with an expander plug-in, so evaluate that too. Experiment with the amount of reverb applied, and also the room size and reverb time, to find a sound that is big but also controlled in the playback of a full kit.

Sidechain processing is often applied with compression, by causing the compressor to respond to a different audio source than the one it is processing. Conventionally, a compressor looks at the signal level to detect threshold exceedances and processes the same signal, but it is possible to have the compressor detect with one signal and process another via the sidechain setting. Many gates and expanders have sidechain features too. With this approach, it's possible to tie the dynamics of two signals together, to achieve a more locked or time-aligned sound. For example, a kick drum waveform can be used to pulse the volume of a bass guitar track, making them feel glued together in time. Similarly, sidechaining can be used to attenuate (or *duck*) one sound when another is present. This is commonly used on the radio so that when a presenter speaks, the background music reduces automatically in level, but the ducking effect can also be employed with good results in music production. For example, it is possible to cause an instantaneous drop in a synthesizer track every time a snare drum hits, giving the snare more cut-through and presence in the mix. Sidechaining can also be used to engage an equaliser only when a signal threshold is exceeded, which is exactly how the vocal *de-esser* plug-in is designed, cutting "ss" and "tt" sounds whenever the

Figure 17.6 Sidechain triggering of a 55-Hz sine wave from the kick drum audio channel.

TRY FOR YOURSELF: SIDECHAIN CONTROL

Use a gate or expander sidechain set-up to trigger a 55-Hz sine wave from a kick drum track, as shown in Figure 17.6. To do this, you can add a sine wave oscillator to a new audio channel and then add a gate/expander plug-in to the same channel. Set the gate or expander plug-in's sidechain input to be that of the kick drum recording. Now experiment with the threshold, attack, and release times, and see how you can enhance the tone and power of the kick drum.

Experiment with compression sidechain set-ups too. A good example is to apply a compressor to a synth, piano, or guitar sub-mix, and set the compressor's sidechain to read from the snare drum track. You can now modify the compressor settings to allow the snare drum to push through the mix a little stronger whenever it hits.

---

associated frequency range becomes loud in a particular signal. Sidechaining is used extensively in electronic music because it can generate a powerful and rhythmic pulsing to the music, and this can be of value in other music genres too, though perhaps used at less extreme levels. A good example of sidechaining in electronic or modern indie music is to use a recorded kick drum sound to pulse a low-frequency sine wave in time with the kick drum pattern, giving a musical and powerful sub-frequency to the mix.

17.6 Drum replacement

In the early 2000s, the desire for contemporary drums to sound bigger than life or hyper-real had become the norm. Steven Slate capitalised on this demand by producing a CD of well-recorded and hyper-real drum samples, which could be used to replace and enhance those that had been recorded on a tracking session. By perfectly timing a trigger of a powerful kick or snare sample alongside the recorded drum hits, an incredible and impactful drumbeat could be created in the mix. Steven Slate's drum sample CD was a big success in his home city Los Angeles, so he developed his idea further into the Trigger drum replacement software system (as shown in Figure 17.7). The software works by allowing a user to set thresholds for detecting particular drum sounds, and then using threshold exceedances to play back additional drum samples in time with the original drum hits.

Nowadays, producers of many music genres replace or supplement the recorded drum sounds with drum samples to make them sound larger than life and to improve the control of sounds in the mix. Improvements can be made owing to the fact that pre-recorded drum samples generally do not have any issues with microphone bleed. It's also possible to record drums with trigger sensors attached to the kick and snare if you know that replacement samples

Figure 17.7 Trigger 2 drum replacement software.

will be used in mixdown – which is a very common approach for metal music production. Indeed, Sylvia Massy endorses the use of drum replacement when required:

> If the kick and snare are weaker than I want, there are many ways to adjust and improve the sound. The easiest way is to augment the kick and snare tracks with carefully chosen samples. The samples are tucked in behind the original tracks to reinforce the sound of the kick, snare and toms. A carefully chosen snare sample can even-out inconsistent playing, or fix a drum that had a tired head.[2]

It's certainly possible to record great sounding drums and mix them to a very high standard without the use of drum replacement. For many genres, drum replacement isn't necessary or appropriate, though in others it is possible to use tools such as Slate Trigger to give a subtle but valuable enhancement to drum sounds, if they have not been recorded to the standard or style that was hoped.

17.7 The final mixdown

Finishing a mix is one of the most difficult challenges to an artist, producer, or mix engineer.[3] How do you know when it's finished? Well, many mix engineers will agree that a mix is never really "finished", it's never quite perfect, or at least not to everyone's taste, and that is one of the beauties of creativity. It's important to "own" your mix and be truthful that you have given everything you can and left no stone unturned in the quest for achieving the best possible result. If you have given your best and most-focused attention to the mix, then it has to be good enough. You may reflect on it in the future and decide on new approaches or initiatives to try in the next project, and in many ways our mixes define our progress as creative artists as we move along with changes in fashion and culture. There's always something that could have been done differently or something to improve, but you must celebrate its values and be happy with your achievements. Much like writing and finishing a book!

We've taken a long journey in these chapters and combined many topics, theories, and techniques. We've seen that drum sound brings together a number of technical and artistic concepts, from the fundamental science of drumhead vibration to the practical skills of drum tuning and the creative art of drum recording and production. Whether for improving performance sound on the stage, in the recording studio, or during the mixdown, absorbing the information and reflecting on this book's discussions in the context of your own drum kit, your own tools, your own sonic preferences, and your own preferred music genres is paramount. With that approach, you cannot fail to become an advanced and proficient artist of drum sound, and a true master of bridging musical science and creativity!

Notes

1 Interview with mix engineer Simon Gogerly conducted on 06/10/2020.
2 Interview with producer/engineer Sylvia Massy conducted on 30/08/2020.
3 As discussed in detail by Grammy-winning producers in The dreaded mix sign-off: handing over to mastering, in: *Perspectives on Music Production: Mixing Music*, Routledge, 2017, pp. 257–69, by Rob Toulson.

Index

Note: **Bold** page numbers refer to tables; *italic* page numbers refer to figures and page numbers followed by "n" denote endnotes.

For Product Safety Concerns and Information please contact our EU
representative GPSR@taylorandfrancis.com
Taylor & Francis Verlag GmbH, Kaufingerstraße 24, 80331 München, Germany

www.ingramcontent.com/pod-product-compliance
Ingram Content Group UK Ltd.
Pitfield, Milton Keynes, MK11 3LW, UK
UKHW020933180425
457613UK00013B/340